The Science of Navigation

The Science
of Navigation

FROM DEAD RECKONING TO GPS

Mark Denny

THE JOHNS HOPKINS UNIVERSITY PRESS
BALTIMORE

The Johns Hopkins University Press
2715 North Charles Street
Baltimore, Maryland 21218-4363
www.press.jhu.edu

Library of Congress Cataloging-in-Publication Data

Denny, Mark, 1953–
    The science of navigation : from dead reckoning to GPS / Mark Denny.
        p. cm.
    Includes bibliographical references and index.
    ISBN-13: 978-1-4214-0511-7 (hdbk. : alk. paper)
    ISBN-13: 978-1-4214-0512-4 (pbk. : alk. paper)
    ISBN-13: 978-1-4214-0560-5 (electronic)
    ISBN-10: 1-4214-0511-3 (hdbk. : alk. paper)
    ISBN-10: 1-4214-0512-1 (pbk. : alk. paper)
    ISBN-10: 1-4214-0560-1 (electronic)
    1. Navigation—History. 2. Maps. 3. Cartography. 4. Geodesy. I. Title.
    VK559.D44 2012
    623.8909—dc23
                                                            2011037777

A catalog record for this book is available from the British Library.

Special discounts are available for bulk purchases of this book.
For more information, please contact Special Sales at 410-516-6936 or
specialsales@press.jhu.edu.

The Johns Hopkins University Press uses environmentally friendly
book materials, including recycled text paper that is composed of at
least 30 percent post-consumer waste, whenever possible.

Contents

Acknowledgments

Thanks to Steve, of Coastal Instrument and Supply, for permission to reproduce the inset photo of figure 3.16; to Carlos Furuti, for redrawing figure 4.12 for me; to Waldemar J. Poerner, for permission to reproduce the images of figures 5.4, 6.5, and the "briny deep" photo of chapter 5. Thanks also to the staff at the Johns Hopkins University Press for their friendly professionalism, and to Carolyn Moser for (as always) thorough and thoughtful copyediting.

The Science of Navigation

Point of Departure

From rural western Canada to New York City, it didn't occur to me that the airplanes flying my wife Jane and me across the continent applied the latest in navigational technology, really an accumulation of millennia of knowledge. Nowadays, traveling over a significant fraction of the earth's surface involves no great mental or physical effort on our part. We pay our fares, take our seats, and ruminate more on the inconveniences imposed by Homeland Security than on the marvels of flight or, the subject of this book, of navigation. The latter struck home on a much shorter journey, when we needed to reach a small art gallery in Brooklyn from midtown Manhattan. Perhaps it is not surprising that we were completely incapable of the feat. We were gawping up at skyscrapers instead of the more familiar red cedars of home and were overwhelmed by the bustle and vibrancy of the big city, a classic illustration of two hayseeds in from the sticks. More surprisingly, our cabbie also did not know his way around; he turned to his dashboard GPS system to flawlessly guide us.

Three thousand years ago, humans developed the ability to move long distances by sea—and the need to get back home again. We as a species were no longer simply wandering aimlessly from place to place, gathering food as we went, and moving on when we had exhausted the local supply. Three thousand years ago, during the Bronze Age, we had well-established cities and ships and trading structures. At first, we tentatively felt our way along familiar coastlines. A few—I will tell some of their stories—ventured farther afield and were able to find their way back home. These epic journeys formed the stuff of legends, and yet today they would take only a few hours and involve negotiating nothing more hazardous than a chicken sandwich.

Navigation can be defined as the science (perhaps initially an art) of maneuvering safely and efficiently from one part of the world to another. Originally the word referred to ships,[1] including steering and the setting of sails—all the skills needed to get a ship from home port to destination. This book is about our three-thousand-year-long journey from navigational infancy (setting out from a harbor in a sailing craft, navigating by the Pole Star and landmarks) to maturity (setting out from Times Square in a cab, navigating by GPS). I describe not only the astounding feats of seamanship and of determination in adverse circumstances of the early explorers, sailors, and conquerors, but also the technical advances made by them, and by many others behind the scenes, who developed the science of navigation.

My intent is to provide you with an understanding of the physical principles that are needed to appreciate navigation and how they came to be understood. I hope that you will close this book with a deeper appreciation of the physical, engineering, and mathematical ideas that have contributed to developments in these fields and will see several layers beneath the surface of modern navigational techniques. In these pages I survey a large area—covering a lot of ground. Continuing the metaphor: the map I construct is large scale, so it necessarily omits much detail. Thus, for example, the fascinating, lengthy, and multifaceted story of the Great Trigonometric Survey of India is here compressed into a few pages. Another example: the well-documented and historically important survey of England is omitted entirely. If you crave details, then please consult the extensive bibliography. (Had I attempted a comprehensive coverage of all the disciplines discussed in these pages, this book would have required 10 volumes. More: 100 volumes. Under a single cover I can only explain the fundamental, overarching ideas and hope to whet your appetite.)

The first chapters encompass preliminary material we will need later on: geodesy and cartography. They set the table. These subjects, important in their own right, are the appetizers before the main course of navigation (apologies for the pun). I show you how navigation evolved and, along the way, tell the stories of many historically important explorers who employed the navigational tools and techniques I discuss.

In chronological order our explorers are the unknown Phoenicians who ventured into the southern hemisphere; two Carthaginians, Hanno and Himilco, who expanded their trading contacts beyond the Mediterranean and along the Atlantic coast; Pytheas, the Greek who ventured to the Arctic Circle; Zhang Qian, the Chinese who explored the eastern lands of Alex-

---

1. The word *navigate* is derived from Latin *navis*, "ship," and *agere*, "to drive."

ander the Great; the South Pacific explorers; Ibn Battuta; Bartolomeu Dias, Columbus and Vespucci, Vasco da Gama, and Magellan; Sir Francis Drake; James Cook; Nain Singh Rawat. These men cover 2,500 years of history and make their appearance in the book at just the right place—very appropriately for explorers.

When we reach the end of the book, we will find that we have been preceded by bats, dolphins, cuckoos, warblers, and other animals that evolved extremely impressive navigational skills long before humans arrived on the world stage.

There are plenty of books which detail the astonishing exploratory journeys that have been made over the centuries (many of the better books are included in the bibliography), and there are many texts aimed at students of the various fields we will touch upon here. Geodesy, cartography, and remote sensing form the technical backbone of our subject: all of these disciplines are intimately connected with navigation. There are remarkably few books, however, in which the authors seek as I do to explain the technical aspects of navigation without relying heavily upon mathematics. How to convey such technical understanding without math? Navigation is, at its core, a geometrical science, and so many of its principles can be conveyed very effectively with diagrams. I have gone to some considerable effort to create pictorial illustrations of the mathematical ideas underpinning navigation, and these diagrams along with some helpful text will take the place of pages of convoluted mathematical equations. Nitty-gritty details (of how to measure or calculate your position on the surface of the globe) are not a part of the story I will be telling, but you will learn why it all works.

Geodesy, cartography, remote sensing—all of these disciplines are interwoven with navigation because, after all, navigators plot a course on charts of the world, or a part of the world, based upon measurements that they have made. People from very different cultures and historical periods have contributed to the enormous corpus of knowledge that is necessary for all aspects of navigation. These people will have had very different outlooks and motivations, but they faced the same physical problem: how to get from A to B. Their knowledge and insights were recorded and passed down to the next generation—in some cases, to the next hundred generations.

Geodesy is explored in the early chapters because, of course, we need to know the nature of the object—the beautiful, ugly, harsh, and beatific spheroid upon which we walk, sail, and pass our lives as it hurtles through space. Cartography—mapmaking—is the handmaiden of navigation. Over the centuries, maps have helped guide ships, and ships have navigated to unknown parts of the world in order to map them. Some of the tools and methods of cartography are carried over directly into navigation. So, cartography is mapped in some of the chapters to follow. Remote sensing is the essence of navigational data gathering. The term today applies to our amazing electronic tools—radar and sonar, optical and infrared cameras, magnetometers, and other measuring devices—that are used on, under, or over both land and sea, and that look down upon the earth from space. It also includes older and more familiar equipment such as telescopes, sextants, and compasses. Navigation has provided the main motivation for developing many of our remote sensing tools, and here I will look at these tools and show you how they have changed the science of navigation.

Although most of the tangential details related to this huge subject have been omitted, I could not resist including some of the more interesting digressions, in boxes separated from the main text so as not to interrupt the flow. So, where I feel that a biographical sketch is called for, or that a navigational technique should be explained in more detail, you will find these scattered throughout the text like islands in the ocean.

The prerequisites you need in order to appreciate this book are (1) a sense of wonder, (2) an interest in navigation and its history, and (3) high school geometry. My geometrical approach is not lightweight, but it is more digestible than algebra. You will be getting the whole meal deal here: real explanations that distill the essence of a problem without belaboring the math. In most of my books I generally resort to mathematical analysis because math is the only language that we scientists share with Mother Nature. Here, however, the subject is so geometrical that I can get a lot of the way there by substituting carefully drawn diagrams and asking you to use your intelligence.

In a sense, navigation is a victim of its own success, in that nowadays we barely need to think about it. If we are lost in a forest wilderness, we do not fear goblins or even exposure; we just text our GPS coordinates to the outside world. As a species, we have learned not to fear falling off the edge of the world or entering unknown realms where there be dragons.[2] We have explored and mapped the entire surface of our planet, above and below the sea surface. We know the precise location of every city and mountain on earth, and we have learned how to communicate rapidly with

---

2. Some old mapmakers labeled far-flung and unknown parts of the world with the phrase "here be dragons."

someone on the other side of the world, and how to get there safely and efficiently. We have learned these things over millennia, and the distillation of that knowledge is something worth knowing. I believe that, after reading this book, you will have an enhanced appreciation for the ingenuity, skill, and achievements of navigators, past and present.

CHAPTER ONE

Earth and Its Orbit

Our subject is navigating over the surface of the earth, and so, to set the table, I begin with some basic facts about our planet and about the influence of its nearest neighbors on it.

Getting to Know Your Neighborhood

Looking at the solar system from a spaceship, far away, you can easily pick out your planet—it's the small blue-green one (volume: about a million million cubic kilometers), third from the sun. In case you do not know what it looks like (thanks to modern remote sensing, you almost certainly do know), a somewhat closer view of the earth is provided in figure 1.1. The colors, owing to the presence of water and life, are what distinguish Earth from the other planets. These two features have also sculpted the earth's surface over eons into the hugely varied landscapes and seascapes that we see today. Suppose that your spaceship is positioned in space so that you see the sun rotating clockwise about its axis. Almost all of the planets, satellites, asteroids, and comets that orbit the sun move in a clockwise manner. That is, you will see the planets orbiting clockwise about the sun and each satellite, such as our moon, orbiting clockwise about its planet. Also, most of the planets rotate about their axis in the same clockwise direction.

Of course, the common direction of rotation is due to the way that the solar system formed, about 4,568 million years ago.[1] Over the eons, the axis of the earth's orbit has changed direction, in a regular and predictable

---

1. The quoted age of the solar system is a recent estimate; see Bouvier and Wadhwa (2010).

way—a phenomenon known as *precession*. The rotational axis direction is not the same as the orbital axis direction. This difference, which is the inclination of the equator to the orbit, is about 23½°, and it is responsible for the seasons. The direction of the inclination changes slowly, a fact that has consequences for navigation: Polaris, the Pole Star, is not always due north. The shape of our planet's orbit about the sun (indeed, as Sir Isaac Newton famously showed us, of all planets) is an ellipse.[2] For Earth, the ellipse is very nearly a circle: the furthest and nearest distances to the sun (*aphelion* and *perihelion*) differ from the average distance by only 1.6%. The earth is not quite a sphere, though for most purposes we can approximate it as one. In fact, the distance from the North Pole or the South Pole to the center is about 6,353 km, and from a point on the equator to the center is 6,384 km (a shade short of 4,000 mi). The equatorial bulge of the earth—due, of course, to the planet's rotation—was not known or understood until the eighteenth century, as we will see.[3]

Your planet has a satellite, the fifth largest in the solar system. By one measure, however, our moon is way bigger than any other satellite: as a fraction of the size of the planet it orbits.[4] This fact has consequences for the exploration of the earth, especially in the days when exploration was carried out in ships: the moon creates large and variable tides. Your planet is dense, like the other inner planets of the solar system but unlike the outer planets. The high density is due to the metal core, which is mostly composed of iron, with some nickel. (Core material is denser near the center of the earth for the same reason as inner planets are denser than outer planets: gravitational pull.) The earth's core is solid from the center out to a radius of about 1,200 km and liquid from 1,200 km out to about 3,400 km. Temperature increases from the surface of the earth toward the center, which is why the outer core is molten iron, not solid iron. (The inner core is solid, despite being hotter, because of the enormous pressure that arises from gravity.) The existence of a liquid iron outer core has consequences for navigation: Earth possesses a magnetic field.

2. In fact, Newton showed that the orbit is an exact ellipse only for a solar system with one planet. The existence of other planets makes the analysis of planetary orbits, and the orbits themselves, much more complicated. However, it is a very good approximation to say that the orbit of each planet is an ellipse.

3. Earth orbital and dimensional data is available from many sources, such as Weast (1973).

4. Not counting Charon, the main satellite of Pluto. Pluto is no longer classified as a planet but is now considered to be a *dwarf plant*, one of five so far identified that orbit our sun.

FIGURE 1.1. Our planet. This famous photograph of the earth was taken from *Apollo 17* while on its way to the moon, in December 1972. Photo courtesy of NASA.

The outer surface of the earth is the least dense. This crust is only a few tens of kilometers thick; it is separated from the outer core by the mantle—2,800 km of hot rock. The crust, especially, is not uniform, and this inhomogeneity results in magnetic (and gravitational) anomalies. The dynamics of the liquid core create interesting movements of the magnetic poles: these effects have significant consequences for navigation.

This brief overview of the earth's structure, and of its orbit, serves to show why we expect that traditional celestial and compass navigation will be affected by the physics of our neighborhood: astrophysics, planetary physics, geophysics. In this chapter I will expand upon the consequences for navigation, and for the closely-related disciplines of geodesy and cartography, of our neighborhood astrophysics and planetary science.

There Is a Tide in the Affairs of Men

The moon orbits the earth at an average speed of about 1 km s^{-1} and takes approximately 27 days, 7 hours, 43 minutes, and 6 seconds to make one

FIGURE 1.2. Earth and moon. The moon exerts a considerable gravitational influence upon the earth, generating tides. This picture was taken by the satellite *Galileo* in 1990. Photo courtesy of NASA.

complete revolution. It rotates, as we have seen, in the same way as the earth; and this means that it takes somewhat longer than 24 hours to reappear in a given part of the sky. In fact, if you look at the moon at 9:00 p.m. one evening, it will again pass through the same part of the sky at 9:50 p.m. the following evening. Our relatively large and dense satellite exerts a considerable gravitational pull, and so the direction of this pull also rotates about the earth once every 24 hours and 50 minutes. Because the oceans are mostly fluid, they flow under the force of lunar gravity: hence, tides. The earth-moon system is shown in figure 1.2.[5]

In most parts of the world, the tides occur twice a day, with a period that is half that of the moon, so that they peak every 12 hours and 25 minutes. Why half? The simple explanation is shown in figure 1.3. The effect of the lunar gravitational force is to pull the surface of the earth that is nearest to it with a greater force than it pulls the center, and to pull the center with a greater force than it pulls the far side of the earth. Thus, relative to the center of the earth, that part of the earth's surface nearest the moon feels a net attraction, and the part furthest away feels a net repulsion. The fluid oceans are able to flow where the lunar force directs them; as a result, the oceans bulge on the side nearest the moon and on the opposite side (fig. 1.3). Another way of understanding this "double bulge" is to con-

5. Lunar and tide data are widely available. See, for example, Weast (1973). For the mechanics of lunar orbit, earth rotation, and tides, and for the data discussed in this section, see Kibble and Berkshire (1985, chap. 6), Lambeck (1980), and Williams (1997). The National Oceanic and Atmospheric Administration (NOAA) also provides much useful information about tides and the causes of tides on its website.

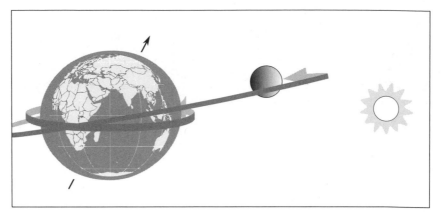

FIGURE 1.3. Moon, sun, and tides. Lunar gravity causes the world's oceans to bulge on the far side as well as the near side, so that tides usually peak twice per day. The sun's gravity also contributes to the tides, though about half as much as the moon's; as a result, tides are particularly high when sun and moon align, as shown here.

sider the centrifugal force that acts on both earth and moon as they spin around one another, like two skaters holding hands. The centrifugal force throws the oceans outward, away from the moon; on the near side, gravity overrides this centrifugal repulsion and pulls the oceans inward, toward the moon.[6]

The double bulge immediately explains why high tides occur twice a day (approximately) instead of once per day. From the point of view of the Man on the Moon, the earth's oceans are shaped like an egg, with the long axis pointing toward him, and with the solid earth rotating beneath these oceans.

This is the standard explanation for one of the prominent features of the tides, but it is by no means the whole story. There are many additional factors that influence the tides. The most important of the celestial influences, apart from the moon, is the sun. The effect of solar gravity on our oceans is about half that of lunar gravity, but the period is slightly different:

6. These two views—of gravity relative to the center of the earth and of gravity plus centrifugal force—are exactly the same to a physicist but are expressed from different frames of reference. The tidal effect of the moon's gravity is also felt by the solid earth as well as by the oceans. However, the solid parts of our planet cannot flow so easily as the oceans, though the earth's crust is distorted 10–50 cm as a consequence of lunar gravity. Undoubtedly it is this distortion that is responsible for the strong correlation between earthquake events and lunar phase: at a given latitude and longitude, earthquakes happen much more commonly when the tidal force is strong than when it is weak.

12 hours instead of 12 hours, 25 minutes. The different positions in the sky of the sun and the moon mean that their combined influence on the tide changes with time. When sun and moon are aligned, as in figure 1.3, the tides are particularly strong. Such unusually high tides occur twice per lunar cycle, when there is a full moon and a new moon, and are known as *spring tides*. In between these times, at quarter and three-quarter moons (when the moon appears as half a circle), the tides are particularly weak— *neap tides*.

Thus, two components of celestial gravity act on our tides. The interplay between them—due to their changing direction and their different periods —makes the behavior of our tides quite complex. Unfortunately, the situation is even more complicated than I have stated it to be. The moon's orbit is not in the same plane as the earth's: it is inclined at 5.145° to the *ecliptic plane* (the plane of the earth's orbit about the sun). Also, the moon's orbit is not quite circular; the shortest distance between the earth and its satellite is about 364,000 km, and the longest distance is about 407,000 km. The difference is big enough to see: at *perigee* (closest distance) the moon looks noticeably larger than it does at *apogee* (furthest distance). Needless to say, the gravitational force is significantly stronger at perigee than at apogee.

The situation is yet more complicated, because the tides are also influenced by factors closer to home. For sure, the tides arise because of the earth's rotation and the gravitational force of moon and sun, but the time and the height of the tides at a given location on the earth are also greatly influenced by local geography, both above and below sea level. Coastline shape, the topography of the ocean floors,[7] ocean currents, wind speed and direction, water flow from estuaries, and atmospheric pressure all influence the height of a tide and the time (relative to the phases of the moon) when it peaks. The result is that each harbor, river mouth, or stretch of coastline has its own tidal pattern, some with large changes in water level due to tides, and some with virtually no tidal fluctuations. Sometimes the *semidiurnal* (twice-daily) nature of tides is masked by other factors so that the tides cycle only once per day or not at all. The types of tide that occur on the coastlines of the world are shown in figure 1.4.

Thus, while most coastlines do experience twice-daily tides as the theory says, in practice the details of tide height and precise timing depend more on local factors. This makes detailed tide predictions difficult. An

---

7. Topography is the study of the shape of land; strictly speaking, here I should say *bathymetry*, which is the study of ocean depths.

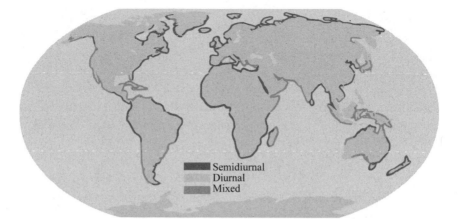

FIGURE 1.4. Types of tides around the world. Semidiurnal tides occur twice daily, with more or less equal amplitude. Diurnal tides happen once per day; mixed tides occur twice a day but with markedly different amplitudes. Adapted from an image by Michael Pidwirny, University of British Columbia.

easy-to-understand example of the influence of local factors is the funneling effect caused by a large bay that narrows significantly: we can expect that such a configuration of coastline might lead to high tides, and this is certainly the case for the Bay of Fundy, between the provinces of Nova Scotia and New Brunswick on the east coast of Canada (fig. 1.5). The tides there are as high as 16 m (53 ft), and these may be the largest tidal fluctuations anywhere in the world—perhaps 5 or 10 times larger than the average.[8] Despite the variability of tide heights around the world, once data has been gathered for a given coastal area it is possible to calculate quite accurately what the future tides in that region will be like. Data is obtained from thousands of tide gauges—measuring stations—installed at harbors all over the world. Predictions are made of high and low tide times and water levels, and are published weeks in advance for the benefit of mariners, fishermen, and other coastal workers (fig. 1.6).[9]

Our understanding of tides, and our ability to predict them in all parts of the world, has increased dramatically over the centuries. Pytheas, the ancient Greek navigator whom we meet later, is said to have recorded the

8. The tidal flow in and out of the Bay of Fundy amounts to a hundred billion tons of water, twice daily. The sheer volume of tides is enough to depress coastlines an average of 15–20 cm.

9. See, for example, Maloney (2006, chap. 17) for the use of tide tables by mariners and for the influence of tides on maritime navigation accuracy.

correlation between spring tides and the phases of the moon. Writers from classical antiquity such as Pliny the Elder and Strabo have much to say on the subject; there are records of tide tables from eleventh-century China and thirteenth-century England. The first mathematical analysis was from the pen of (no surprise) Isaac Newton in the second edition of his famous *Principia*. Newton's analysis was basically correct but missed a number of

FIGURE 1.5. The Bay of Fundy in eastern Canada, at high tide (*top*) and low tide. These are probably the highest tides of the world, exceeding 50 feet at the top of the bay.
Images captured by NASA's *Terra* satellite in April 2001 and September 2002.

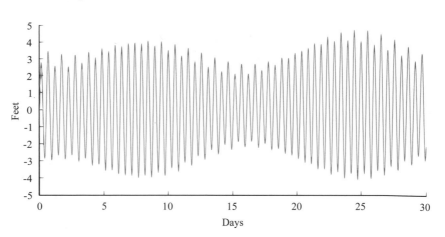

FIGURE 1.6. Calculated tide height in feet, for Bridgeport, Connecticut, September 1991. The variation between neap and spring tides is clear.

important facets. Laplace, later in the eighteenth century, truly nailed the phenomenon of tides mathematically—his equations are still used today.

Modern analysis takes into account a lot more than could Laplace, of course: computers can apply the math to accurate modern data that account for the depth and shape of the oceans, unknown in Laplace's day. They can obtain numerical solutions that show detailed features of worldwide tidal movements which physicists of earlier times could only dream about. Such detailed analyses reveal the presence and shifting location of *amphidromes*; these are points about which the tides oscillate and at which tidal fluctuations are zero. They are like the eye of a storm, in that they are calm areas with large tidal movements swirling about them on all sides, in either a clockwise or counterclockwise direction, depending on the action of *Coriolis forces* and local bathymetry.[10]

Tidal energy dissipates mostly as heat. That is, friction between the moving water and ocean beds and shorelines dissipates much of the energy possessed by tides. The degree of dissipation varies over the oceans, as can

10. More accurately, amphidromes are nodes in a two-dimensional standing wave, analogous to the stationary point at the center of an oscillating string. There are long ocean waves generated by tides that resonate at different dominant wavelengths in different bodies of water. These resonances are sustained by energy that is sapped from the moon; as we will soon see. Coriolis force arises from the earth's rotation; it is a sibling of the centrifugal force but is more complicated. For example, it depends on the velocity of the mass that it acts on.

be seen from figure 1.7. In some parts of the world, such as the North Atlantic, energy dissipation is very large, as much as 30 kW km^{-2}; elsewhere, it can be negative—meaning that parts of the ocean *absorb* energy. The net flow, over all the earth's oceans, is outward: energy is lost as heat. Analysis reveals that the source of this energy is the moon. Thus, the force of lunar gravity moves large bodies of water; the energy required for these tidal movements reduces the gravitational potential energy of the moon (actually, of the earth-moon system), which causes it to drift away from the earth at an rate of about 3.8 cm yr^{-1}. Tidal friction applies a brake to the earth's rotation, so that the spin of our planet is slowing down. This means that the length of an earth day is increasing slowly; over the last century, the length of a day has increased by 2.3 ms. Such a tiny increase may be entirely unnoticed by a person during her lifetime, but over geological timescales the change is significant. Geological data show that, 620 million years ago, one earth day lasted for about 22 hours, and there were approximately 400 days in a year.

It was the astronomer Edmond Halley, of comet fame, who first noticed the energy dissipation phenomenon. Comparing new and old data, he saw a change in the moon's speed across the sky. In fact, what he had noticed was due to the slowing of the earth's rotation speed. The full explanation had to wait until the nineteenth century. Today, we understand the major effects—and a myriad of minor effects—of tidal movements very well. Observations are exquisitely accurate and show, for example, that tides cause the rotation rate of our planet to vary from day to day. This variation is only a few tens of microseconds but is well within the accuracy of our instruments to measure. Such minor details have no consequences for

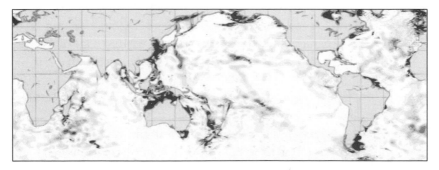

FIGURE 1.7. Tidewater energy dissipation. The darker areas indicate regions of higher energy flux (mostly dissipation or loss of energy from the oceans). Image by Richard Ray, NASA Goddard Space Flight Center.

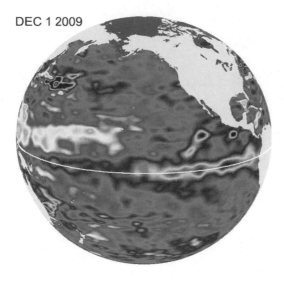

DEC 1 2009

FIGURE 1.8. Satellite altimeter data showing variations in sea level across the Pacific Ocean in December 2009. Warm water is higher than cool water. Image from NASA/JPL Ocean Surface Topography Team.

navigation, but our deep understanding of the physics of tides is very significant because it produces accurate predictions in the form of tide tables.

The influence of solar heating on the flow of seawater is apparent from figure 1.8. This satellite altimeter image shows the variations in sea level height across the Pacific Ocean. Water expands as it warms, and this thermal expansion is responsible for height differences; water flows downhill, and so solar heating influences currents (see "Ocean Currents"). These influences may be seasonal and periodic.

On the Surface

To a good first approximation, the earth is spherical—a ball. More accurately, it is slightly squashed—an ellipsoid. If we seek greater precision of expression, as geographers do, we must specify what we mean by "the earth." The natural (to a physicist) definition of the earth's shape is the *geoid*, which is specified not by the solid or liquid surface of our planet, but by its gravitational contours. In this section, I will explain these increasingly accurate perceptions of the earth's shape.

We will see that humans learned from observations, very early in history, that their planet was round like a ball. From then until a few hundred years ago, we thought of the earth as a sphere, and this approximation is good enough even today for many purposes. In the seventeenth century, Newton realized that it *is* an approximation: the true shape must be a little flattened because of the centrifugal force that arises from rotation. The planet must bulge slightly at the equator and be flattened a little at the poles. On the basis of his law of gravity and an assumption that the earth is a fluid, he calculated that the bulge was about $\frac{1}{230}$; that is, the earth's diameter at the equator is larger than the diameter between the poles by this fraction. In fact, we now know that the earth is not really (or not wholly) a fluid, and so the deviation of its shape from a sphere is less than Newton calculated: it is about $\frac{1}{298}$. (The bulging effect of the centrifugal force is less pronounced than it would be for a fluid because solid material is more resistant to deformation.) A century after Newton, two French expeditions were sent out to different parts of the world to see if his idea about the bulging earth was correct. A controversy raged at the time concerning the shape of the earth, with some eminent French physicists opining that our planet was a prolate spheroid, like an egg (a stretched sphere, instead of a squashed one). We will meet up with these French expeditions in the next chapter.

In the twentieth century, yet more accuracy was needed. For the purposes of *geodesy*,[11] it was no longer sufficient to say that the earth was an ellipsoid. A better definition came by considering gravitational contours. Imagine a large number of plumb lines, all over the earth, the oceans as well as the land—let us say one plumb line on every square meter of the planet. Imagine now a surface that is perpendicular to all those plumb lines. Physicists call this an *equipotential surface*. There are many equipotential surfaces, but one is special. The geoid is the equipotential surface that coincides with mean (average) sea level. It is considered to be the true shape of the earth.

Clearly, the geoid must be very similar to the earlier ellipsoid, because it pretty much coincides with it over the sea.[12] There are slight deviations, especially over land, due to the uneven distribution of mass within the earth. A large deposit of dense metallic ore will cause the geoid to bulge outward

11. According to the *Oxford English Dictionary*, "That branch of applied mathematics which determines the figures and areas of large portions of the Earth's surface, and the figure of the Earth as a whole."

12. The seas, being fluid, flow if a force acts on them. Consequently, the sea surface must coincide with a gravitational equipotential surface. Think about it (and let's suspend the moon's influence for a moment): if there were a horizontal component of the gravitational force on one part of the sea surface, the sea would move under the action of this force to a position where there was no such horizontal force. That is, the sea would flow until its surface was equipotential (in practice, very close to the geoid). Note that the geoid is not necessarily a surface of equal gravitational *strength* because of the effect of the centrifugal force.

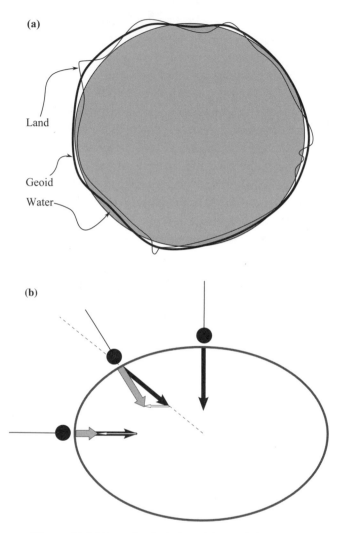

(a)

Land

Geoid

Water

(b)

FIGURE 1.9. The geoid. (a) Here, the deviation of the earth from an ellipsoid is greatly exaggerated. The geoid is an equipotential surface, so it largely overlaps with sea level, with slight variations due to underlying mass irregularities (gravitational anomalies). (b) For an exaggerated ellipsoid, we see why the plumb lines are perpendicular. To the force of gravity (black arrows, acting toward the ellipsoid center) we add centrifugal force (white arrows) resulting in the total force (broad arrows) that determines the plumb line direction. The geoid is the surface that is perpendicular to all possible plumb lines. Note that anomalies (not shown) can cause the plumb line direction to veer; therefore, the geoid is slightly different from the ellipsoid.

at the location of the ore, for example. A less dense area will show up as a local dip. In fact, the geoid is a very complicated shape, differing from the simple ellipsoid of earlier geographers by as much as 110 m in places (i.e., up to 110 m higher or lower than the ellipsoid). The idea of the geoid is illustrated in figure 1.9; here, you can also see a geometrical explanation of why it is close to being an ellipsoid. Figure 1.10 shows an example of local variations in the gravitational field—in this case, over the state of New Jersey.

Modern instruments can measure the force of gravity—and therefore can measure gravitational equipotential surfaces—very accurately indeed. The "breathing" of the earth, as it bulges and contracts due to the changing direction of lunar gravity, can be detected. To define a fixed geoid, this tidal

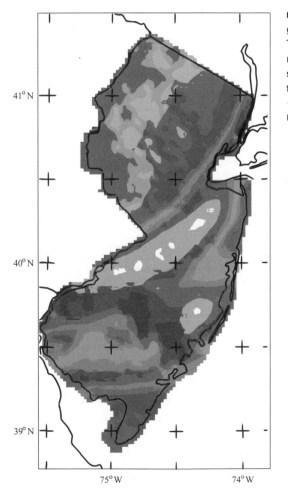

FIGURE 1.10. Local variation in gravitational field intensity. This gravitational anomaly map of New Jersey shows small variations of gravitational acceleration, from $-0.05\,\mathrm{cm\,s^{-2}}$ to $+0.03\,\mathrm{cm\,s^{-2}}$. U.S. Geological Survey.

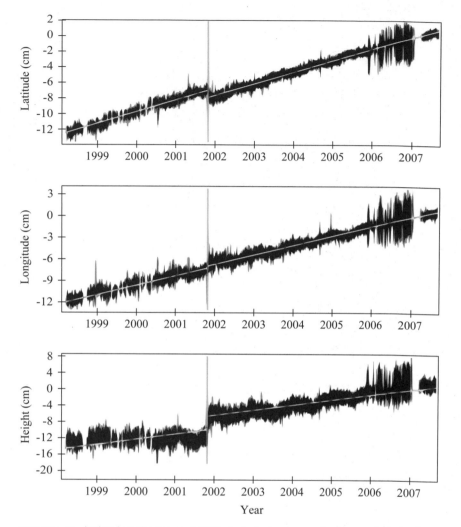

FIGURE 1.11. Iceland on the move. A GPS site in Iceland records changes in latitude, longitude, and height over a period of 8 years. Adapted from a NASA graphic.

effect must be averaged out, and so must the more noticeable ocean tidal movements. (One natural phenomenon that is not averaged, however, is the rising and falling of sea levels, as ice ages come and go and air temperatures cool and warm. Thus, the geoid changes slowly, on a geological timescale.)[13]

13. Many textbooks explain geoids and geodesy, and I will reference a few in chapter 2, where we come to grips with geodesy and how it developed. For this chapter the reader

Thus, the earth's crust is not as solid as we like to think. It is more like an animal's skin than a hard shell; it contracts, distorts, and twitches in response to internal as well as external stimuli. These days even plate tectonics—the slow movement of continents—can be measured accurately, to a degree that was unimaginable even a generation ago. Later I get to tell you about GPS; here I will settle for presenting some of the geodetic data that GPS has provided for us. In figure 1.11 you can see how one site on the surface of the earth has been moving over an eight-year period. Many such sites have been monitored, providing detailed data about the whole surface—the dream of geodesy.

Wandering Star

If you were to draw a line from the South Pole to the North Pole, and then extend that line for some 300 light years, it would pass close to Polaris, a double star (which looks like a single star—the 49th brightest in the sky) in the constellation Ursa Minor. Stated differently, if you were to lie on your back and stare upward into a clear night sky, Polaris is the only star you would observe *not* to follow a circular path, as the earth rotates. Polaris is our North Star; it is close to being due north, directly over the North Pole, which means that any navigator in the northern hemisphere can determine the direction north, on a clear night. The fact that Polaris is the only fixed star in the northern sky is due, of course, to the earth's rotation.

The earth's axis of rotation is not parallel to the axis of its orbit about the sun. The *obliquity angle* of 23½° is approximately constant (it oscillates between 22.1° and 24.5°),[14] but the direction in which it points varies with time: the rotation axis circles around the orbital axis, as shown in fig. 1.12a, once every 25,700 years. The reason for this *lunisolar precession*—analogous to the slow precession of a spinning top or a gyroscope about a vertical axis—does not really concern us, but the fact of its existence does have navigational consequences. Precession means that Polaris is not al-

---

might find a nontechnical introduction to geoids, such as the one at www.esri.com, more useful.

14. This angle is known technically as the *obliquity of the ecliptic*, where, as we have seen, the ecliptic plane is the plane defined by the elliptical orbit of the earth. I try to avoid such technical jargon wherever possible in this book (celestial mechanics is full of it) and employ only those terms that have a significant influence on navigation. Note, incidentally, that the ecliptic plane wobbles slightly—it is not fixed in space—and that the direction of the long axis of the ellipse also moves. These effects are small.

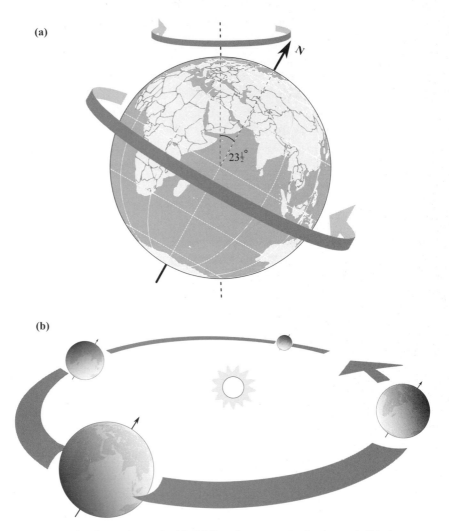

FIGURE 1.12. Earth rotation and orbit. (a) Our planet rotates about an axis (black arrow) that is inclined at an angle of about 23½° to the orbital axis. The rotational axis precesses about the orbital axis, making one rotation every 25,700 years. (b) The inclination angle is responsible for the seasons because the length of day varies with orbital position.

ways the North Star. Today it is three-quarters of a degree away from true north, but it is getting closer. The closest that Polaris will be to our true north is 0° 21′31″ (0 degree, 21 minutes, 31 seconds),[15] and this will occur

15. One minute of arc is ⅟₆₀ of a degree; one second of arc is ⅟₆₀ of a minute. Thank the Babylonians for this sextagesimal system.

in the year 2102 CE. Much later in the future, around the year 7500 CE, the closest star to true north will be Alpha Cephei, not Polaris. In 15000 CE it will be Vega. We will have to wait another 24,000 years or so before Polaris again becomes the North Star. Five thousand years ago, mariners who looked to the North Star would have seen Thuban (Alpha Draconis). For Pytheas, the North Star was Polaris, but it was several degrees further away from true north than it is today.[16]

Here is another aspect of rotation: the obliquity angle is directly responsible for the four seasons that we experience. The rotational and orbital geometry is illustrated in figure 1.12b. From this illustration we see why the seasons arise. At different places in the earth's orbit, the number of hours of daylight is different for any fixed position on the earth's surface (say, your house). In summer you see the sun rise in the northeast (if you're a northern hemisphere resident), move quite high into the sky, and then set in the northwest. This happens when the rotation axis is inclined toward the sun. In winter you see the sun rise in the southeast and cross low in the sky to the southwest, where it sets—and this occurs when the rotation axis is inclined away from the sun.

There are four lines of latitude that are so special that they are given their own names. Their distinctiveness arises because of the obliquity of the ecliptic (the 23½° inclination angle). The Arctic Circle lies at latitude 66½° (that is, 23½° from true north); it marks the southernmost latitude in the northern hemisphere at which the sun can be above the horizon 24 hours a day. Similarly, the Antarctic Circle is the northernmost latitude in the southern hemisphere at which the sun can be above the horizon all day. In other words, you have to be inside the Arctic or Antarctic Circles to experience a midnight sun. The Tropic of Cancer lies at latitude 23½° north and the Tropic of Capricorn at 23½° south. In between these latitudes—in "the tropics"—you can experience the sun directly overhead, vertically above you. Outside the tropics, you cannot.[17]

What if the obliquity were 0°—if the earth's rotation axis were aligned

16. Just to confuse you, I will point out here that the location of true north is also not fixed. *Polar motion*—the movement of the earth's rotation axis, and so of the poles, across the surface—amounts to a few meters per century. Polar motion is caused by movements within the earth's mantle and core and by redistribution of mass following the last ice age. I will have more to say about polar motion in chapter 2.

17. The seasons have nothing to do with earth's distance from the sun. The closest point of approach (perihelion) occurs on or about January 4. Data for this section comes from Burnham (1979, 3:2009). For a nonmathematical explanation of celestial phenomena, see Kaler (1996).

with the orbital axis? In that case we would have no seasons, no variation in day length. The height of the sun in the sky, and the intensity of sunlight, would vary with latitude, but the days would all be the same length at all latitudes, and for every day of the year. At the other extreme, we might ask what the seasons would be like if our planet's rotation axis were tilted over 90° (that of Uranus is actually more than that: it is 97.86°). At the poles, the sun would be below the horizon for nearly six months of the years, during what would surely be a very deep winter, and above the horizon for the remaining time—a broiling summer. At latitude 45° north or south, the sun would remain below the horizon for nearly three months of the year, and for a slightly longer period it would not set. At the equator, the sun would vary in height from directly overhead on midsummer's day to skirting the horizon in midwinter. The world would be a very different place.

Magnetic Mysteries

The Song-dynasty Chinese scientist and statesman Shen Gua wrote about the magnetic compass in 1088 CE; from his and other writings it is clear that the compass had been in use by Chinese mariners for some centuries. The earliest written reference to the compass in Europe dates from the thirteenth century, so it seems likely that the magnetic compass originated in Asia. These compasses made use of *lodestone*, a naturally magnetic material; with lodestone it proved possible to magnetize an iron needle which, when attached to a low-friction bearing (or floated in a little boat in a bowl of water), would orientate itself along the geomagnetic field lines, thus telling a navigator which way is north.[18]

Let's back up a little. Why should there be a geomagnetic field at all? Given that one exists, why is it more or less aligned north-south? The physics of planetary magnetism is very complicated and is still an area of active research. The phenomenon has been well studied for centuries because of its importance for maritime navigation, but an understanding of the underlying causes began only in the 1950s. Basically, the magnetism arises from the liquid metal of the earth's outer core: the turbulent convection of the rotating fluid generates electrical currents (the outer core is a good conductor of electricity) which in turn produce a magnetic field.[19]

18. See Needham (1959, pp. 26–27) or Selin (1997) for more on the Chinese origin of magnetic compasses.

19. Researchers suggest that an initial magnetic field would induce a current in the conducting outer core, and this current would in turn induce a magnetic field. Depending

The largest component of this magnetic field is a dipole (like the field produced by a bar magnet) that is aligned along the earth's rotation axis. There are other, non-dipole components of the field, however, and these (along with irregularities of the core, mantle, and crust) greatly complicate the observed field patterns. Add in the intrinsic complexity of the physics (magnetohydrodynamics), and you can see why a detailed understanding was slow in coming.

The basic model is that of a dynamo—in fact, of two coupled dynamos. The dynamics of such a system are chaotic, just like the earth's magnetic field. The polarity of the twin-dynamo model flips chaotically[20]—that is, the north and south poles suddenly swap positions, and then swap back again in an irregular and never-ending sequence. This is very reminiscent of the well-known reversal of the earth's magnetic field, which has been detected via geological records. It occurs every few hundred thousand years but in a seemingly random fashion: the last reversal was 780,000 years ago.[21] Each reversal takes a few thousand years to complete—a blink of an eye, geologically—and during a reversal there is no reliable field orientation. One feature of the dynamo model is that it can explain key features not only of the chaotic magnetism of our planet, but also of the regular 11-year cycle of magnetic polarity flips that occur in the sun, which is a very different physical system.[22]

on the physical configuration, the induced magnetic field and electrical current can then be self-sustaining—even if the initial magnetic field were to disappear—so long as an energy source exists to maintain convective flow in the outer core. (Perhaps natural radioactivity generates heat that leads to convective flow in the outer core.) The self-excited dynamo model discussed in the text is a simple example of this idea. Our knowledge of the earth's interior is imprecise, however. None of the core's constituent matter can be measured directly, so its nature is inferred. In particular, we do not know much about the conductivity and other magnetohydrodynamic properties of matter at the high temperatures and very high pressures that exist in the earth's interior. See Kearey et al. (2009, pp. 74–77) for a recent accessible account of this complex phenomenon.

20. Chaotic events appear to be random, but they are really a consequence of the underlying dynamics. The equations that govern meteorology are similarly chaotic, which is why detailed, long-range weather forecasting is so hard to do. Predictions can be made a little way into the future, given accurate data of the current state of the weather, but they become worse as we look further into the future. The same situation exists with geomagnetic field predictions.

21. The five previous reversals were at 990,000 years ago and at 1.07 million, 1.19 million, 1.20 million, and 1.77 million years ago. Polarity flip simulations can often be seen online.

22. Technical and semitechnical accounts of the physics of geomagnetism can be found in Buffet (2000), Carrigan and Gubbins (1979), Kibble and Berkshire (1985, p. 265),

In figure 1.13 we see a much-simplified geomagnetic field structure, which nevertheless shows two features that are important to us. First, the field is not aligned exactly north-south. That is, magnetic north is not the same as true north. In fact, it moves. In 2005, the location of magnetic north was approximately at 82° 54′ north, 114° 24′ west. It is moving from northern Canada toward Siberia at a rate of 60 km yr^{-1} and accelerating. The angular difference between magnetic north and true north is called *declination* by geophysicists and *variation* by navigators. It is a great problem for navigation because it means that a compass does not point northward; the error may be small in equatorial regions, but it can be very large near the poles. For this reason, navigators have for centuries measured the magnetic field of the earth so that calculations can be made to compensate for the compass variation error.[23]

The second feature of the geomagnetic field that is apparent in figure 1.13 is the *inclination angle*; this is the angle that the field lines make with the surface of the earth. Near the equator, the inclination is 0°; moving toward the magnetic poles we see that the inclination increases.

We met Edmond Halley earlier in this chapter. He was one of the first scientists to investigate the variation of the earth's magnetic field, and undertook a sea voyage over the North and South Atlantic in order to measure it. His magnetic map was the first to use contour lines to indicate regions of constant magnetic variation (*isogonic* lines). He published his book *New and Correct Chart Shewing the Variations of the Compass* in 1701. Subsequent measurements over the centuries revealed that the variations are changing, and doing so quite significantly. Not only does the location of magnetic north vary, but so does magnetic south—and magnetic south is not even diametrically opposite magnetic north. Also, inclination angles vary with time, for any fixed location on the earth's surface. The intensity as well as the direction of the field varies; it has been weakening for the last century or so. Because of these continuous and substantial changes, it became necessary to establish a worldwide network of stations to monitor the geomagnetic field. Such a network was first set up in the nineteenth century, and today there are 170 magnetic observatories spread across the globe.[24]

Moffatt (1993), and Rikitake (1958). The website of the U.S. Geological Survey (USGS) provides a clear, nontechnical explanation.

23. The problems that magnetic variations cause when one is navigating toward the true North Pole are made plain in Avery (2009).

24. For Halley's measurements of magnetic variation, see Cook (1998, pp. 131, 270–71, 283). Halley also produced tide charts (pp. 288–89).

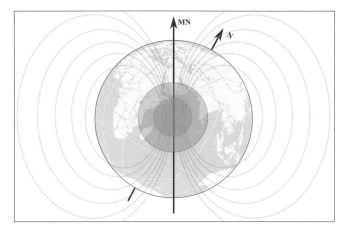

FIGURE 1.13. Idealized geomagnetic field lines. Note the variation of magnetic north from true north, and the inclination angle of the field lines with the earth's surface.

Another source of geomagnetic field variation is the sun. There are from time to time huge eruptions from the solar surface that are known as *coronal mass ejections*; these disturb our planet's own field and occasionally even link directly to it, resulting in large magnetic storms, during which all bets are off regarding field variation predictability. The heating of our ionosphere during these magnetic storms adversely affects navigation in a couple of different ways: long-range radio communications are interrupted, and GPS service is degraded. (Also the aurorae—borealis and australis—become more prominent, and electrical blackouts happen as a result of power surges along national power grids.)

The overall shape of the field is not as simple as figure 1.13 suggests: there are many local anomalies due, for example, to iron ore deposits in the earth's crust, but on a larger scale due to irregularities in the outer core and to non-dipole features of the field. Thus, in the year 2000 the geomagnetic field looked like figure 1.14. From this chart it is clear that real magnetic field lines wander quite far from the idealized north-south direction of figure 1.13.[25] Following a compass from the magnetic South Pole will eventually lead you to the magnetic North Pole, but your path will not be straight. Because every feature of the geomagnetic field changes in time, magnetic maps soon go out of date and need to be replaced every five years. The geophysicists' understanding of the phenomenon is now sufficiently

25. An animation of the change in magnetic variation over the past few centuries is provided online at Wikipedia's entry for magnetic declination.

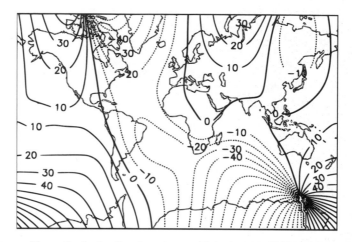

FIGURE 1.14. Magnetic declination from true north in the year 2000. Map from the U.S. Geological Survey.

good, however, that they can predict ahead a few years. Thus, a magnetic chart for 2012 can be predicted with reasonable accuracy from a map obtained in 2010 from measured data.

Winds and Waves

Historically, much of navigation has been in the hands of mariners aboard sailing ships. These ships are subject to the vicissitudes of wind and water; it therefore behooves me to spend a small part of this chapter discussing the patterns of winds and of surface ocean currents. These fluid movements matter because they are long-lasting or seasonal, and predictable, and thus can be a great help to mariners if not to navigators. A book twice the size of this one might get to the heart of meteorology, but a single-section summary can only scratch the surface. Some of the history of navigation and exploration depends on established wind and ocean currents (for example, the discovery of Brazil by Europeans), and so you will need to be aware of the existence of these phenomena, even if their cause and interaction must remain mysterious.

The basic engine of planetary wind patterns is atmospheric heating. This happens more at the equator than anywhere else. As a consequence, equatorial regions tend to be low-pressure areas: heated air rises. This rising air circulates poleward and cools, descending at the so-called *horse latitudes*, which are 25° to 30° north and south. The cooling, falling air

FIGURE 1.15. Hurricane Isabel, September 17, 2003. Image courtesy of NASA; taken from the orbiting International Space Station.

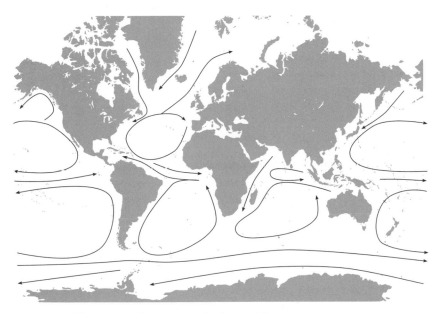

FIGURE 1.16. The main surface currents in the world's oceans.

Ocean Currents

The details of ocean current form, strength, and evolution are complex. These depend on the earth's rotation, the topography of the ocean floors, water salinity and heat content, and the prevailing winds. Much of the detail is not well understood and consequently is not predictable, but certain trends are clear. There is, for example, a global conveyor belt of heat—warm surface water in the tropics that is driven north and south by winds, away from the equator. These water bodies then cool, sink, and drift back toward the equator at depth.

Some currents are strong and permanent. The Gulf Stream is a warm surface current flowing north from Florida to Newfoundland before splitting in mid-Atlantic into a western European stream and a West African stream. This current warms the east coast of North America, and Western Europe. The California current is cool and shallow; it moves south from British Columbia to Baja California. One consequence of this current is the cool waters off Hawaii. The Brazil current was historically important because, during the Age of Exploration, it took Portuguese ships off the west coast of Africa and carried them southward to Brazil—leading to the Europeans' discovery of that vast region. The South Atlantic current then brought them eastward back to southern Africa.

Navigators learned to trust these currents. Even today, currents are important to shipping, if not to navigation, because a judicious choice of route that takes currents into account can significantly reduce transport costs.

---

increases air pressure. This basic circulation pattern is responsible for the *trade winds*: air at the surface flows from regions of high pressure (horse latitudes) to low pressure regions (the equator). They get deflected by the Coriolis force, resulting in winds that are from the northeast in the northern hemisphere and from the southeast in the southern hemisphere. At the equator, the winds are less reliable; they are light and variable, and we have the *doldrums*.[26]

North and south of the horse latitudes, the pattern of high and low

26. The origin of the phrase *horse latitudes* is uncertain; tradition has it that Spanish sailors, on their way to their Caribbean colonies, became becalmed and threw their horses overboard to save on drinking water. *Trade winds* are so called because, being strong and reliable, they greatly aided the passage of merchant ships across oceans. Sailing vessels were becalmed in the *doldrums*; later, the word took on a more general meaning of lethargy.

pressure repeats, resulting in the *polar front* at 60° latitude—regions of low pressure—and *polar caps*, where the pressure is high. These zonal patterns are displaced north and south seasonally, because of the changing amount of solar heat that reaches a given area. Such seasonal winds include the *monsoon* of India and Southeast Asia. *Cyclones* are seasonal; they are circulating areas of low pressure that give rise to *hurricanes* in the North Atlantic (fig. 1.15). The same phenomena are called *typhoons* in the western Pacific. The *roaring forties* are regions between 40° and 50° south, with strong prevailing westerlies that are unimpeded by land masses.

Ocean currents are driven by solar heating, winds, gravity, and the ubiquitous Coriolis force. Water in equatorial regions is heated and so expands, raising the sea level some 8 cm over vast areas of ocean. The water falls down this slope, giving rise to currents that are modified by winds and land masses. A map of the major ocean surface currents is shown in figure 1.16.

Winds and ocean currents have influenced navigation in two ways. First, they drove sailing ships before them, whether or not the pilots wished it; and secondly, they exacerbated the problem of estimating the speed of a sailing ship and so made dead reckoning more difficult. Dead reckoning is discussed in chapter 6, and so I will say no more about it here.

CHAPTER TWO

Shaping the Earth

Here, we look at the perceived shape of the earth and how this perception changed over the centuries. We begin (once a myth has been dispelled) as so many discussions of scientific development must begin, with the ancient Greeks.

Falling Off the Edge of the World

Let's get something out of the way before we start. When I first learned about Christopher Columbus, as a young schoolboy in England 45 years ago, I was told about his mutinous and terrified crew. They fervently believed that the earth was flat, that Columbus was steering his three little ships toward the edge of a disk, and that if he did not order them to turn back soon, they would all fall off the world. I had visions of a gigantic waterfall, with three little medieval wooden sailing ships tipping over it. Only it didn't happen because the great Columbus knew better: he stayed on his westward course, and just before the ignorant mutineers were about to overpower him, land was sighted to the west. America was found, Columbus was vindicated, and everyone lived happily ever after.

Garbage. My childishly naïve view of this history was wrong in every detail, of course. Of interest to us here is the first error: the idea that medieval Europe believed in a flat earth is just plain wrong, as even the most cursory historical research reveals. This myth is a figment of the nineteenth century and persisted, it seems, at least until the middle of the twentieth.[1]

1. The myth of a flat earth is still with us today, despite space travel and the evidence provided by modern technology and common sense. The Flat Earth Society is on the web,

For sure, early cultures believed that the world was essentially flat, with local irregularities. That is, mountains and valleys were superimposed upon a globally flat surface, rather than upon a spherical surface. Homer, writing in the ninth century BCE, saw the world as a circular disk—a plateau surrounded by water, with Hades below and the dome of the sky above. The Chinese maintained until the seventeenth century that the world was flat, but most other peoples had abandoned this view fifteen hundred years before—earlier in the case of classical Greece, as we will see.[2] A few scholars and Christian dogmatists flatly refused to adopt a well-rounded view of the world, but most educated people and all mariners knew different (fig. 2.1).

Before looking at the ancient evidence that led the great majority of educated Europeans to believe that the world was spherical, I will spend a paragraph demonstrating that they did so from quite early on, and certainly in medieval times before the great maritime explorations began. It is simply a matter of consulting ancient writings. Thus, the Venerable Bede, an eighth-century English monk and historian of Dark Age England, makes clear that the earth is a globe, like a ball. Dante Alighieri's masterpiece, *The Divine Comedy*, written sometime during the first quarter of the fourteenth century, portrays the earth as a sphere and takes note of the different stars that are visible from the southern hemisphere. At the end of that century, a hundred years before Columbus set out on his first transatlantic voyage, the English writer Geoffrey Chaucer produced *A Treatise on the Astrolabe* for his son, which very clearly displays his view on the spherical shape of the earth. (Indeed, as we will see, the very existence of the astrolabe is a sharp reminder that the earth is a sphere.) Many other writings, from many countries over several centuries, show the ubiquity of this "modern" view of the world. Modern technology leaves twenty-first-century earthlings in no doubt as to the roundness of their planet (fig. 2.2), but as so many Dark Age and medieval authors show us, such high-tech confirmation is hardly essential.[3]

So, the belief in a flat earth had fallen flat by late antiquity and certainly

should you feel the need to follow up on this bizarre notion. As Boorstin (1983) says, "The greatest obstacle to discovering the shape of the earth, the continents and the oceans was not ignorance but the illusion of knowledge."

2. On the persistence of the Chinese flat-earth view, see Needham (1959).

3. For Bede and Chaucer's writings, see Harris and Grigsby (2008). Chaucer's *Treatise on the Astrolabe* can be read online. Dante's spherical-earth views are discussed in Harley and Woodward (1987).

by early medieval times in Europe. Why the change in view? Observational science came to influence Greek philosophers, and the evidence for a spherical earth was right before their eyes. Pythagoras preferred a spherical earth anyway, on aesthetic grounds (typical for him). Aristotle, who lived two centuries after Pythagoras, placed greater value on evidence, and from these observations he too considered the earth to be a sphere. Strabo, a Greek geographer writing at the turn of the common era, deduced a spherical globe on the basis of what he saw closer to the surface. He noted that lights in elevated lighthouses could be seen from farther out to sea than lights at sea level. Masts of ships disappeared from view after the ship's hull. Such maritime observations—which applied to observers looking in all directions—strongly suggest a curved earth. Eratosthenes, another Greek and a confirmed scientist, not only believed that the earth was a sphere but also measured its radius—for which he, and a couple of other like-minded philosophers, deserve a separate section in this book.

The crew of Christopher Columbus may well have been close to mutiny. The crews of many sailing ships that explored the world at earlier and later

FIGURE 2.1. A fifteenth-century depiction of the earth as a globe. This illustration, from a book by John Gower (written ca. 1400 CE and now in Glasgow University Library), shows him shooting at the world: "I throw my darts and shoot my arrows at the world. But where there is a righteous man, no arrow strikes. But I wound those who live wickedly."

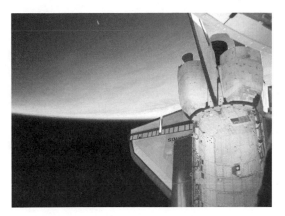

FIGURE 2.2. Our planet seen from space. It is clearly round and not flat; the ancients knew this, even without such direct visual confirmation. Image from NASA.

times—such as those of Pytheas, or of the Portuguese, Spanish, English, and Dutch who so determinedly set their sights on the other side of the known world—may also have come close to rebellion. Magellan and Drake both had to suppress mutinies. The reason, however, was much more mundane than a fear of falling off the edge of the world, or dread of sea dragons or other monsters of the imagination. Lack of food and water and a fear of death amid towering waves understandably cut through the enthusiasm of mariners. Their captains may have been driven by zeal of one sort or another—the desire to spread the word of God, the desire to become very rich, or simply the desire to see what was out there—but the average sailor was understandably more down to earth. He will have been, no doubt, quite happy to become rich, or to see sights never before spied by man, but not at the cost of his life.

The drive of the early explorers—the captains who strived for years to find sponsors for their expeditions, and who persevered relentlessly despite the fears of their crews—is all the more admirable to us, glimpsed centuries later from the comfort of our armchairs, precisely because they were swimming against the tide, so to say. They overcame many obstacles, but the ancient fear of falling off the edge of the world was not one of them.

Measuring the Earth's Radius in Ancient Times

Homer (ninth century BCE) was a flat-earther, as were most people in his times. More particularly, the earth to him was a disk on a plateau that was surrounded by the river Oceanus. After all, if the earth were not flat then we would slide off it. Furthermore it was stationary, because if it moved we would feel the motion. And it was the center of the universe, because the stars rotated around it. Another early, pre-Socratic philosopher, Thales of Miletus, may have considered the earth to be a disk resting on water. We will meet Thales again; he had some more progressive ideas about geodesy. His student Anaximander (early sixth century BCE) thought of the earth as a cylinder, with a height three times its diameter. The axis of this cylinder was oblique to the axis of the sun, and the earth was at the center of a celestial sphere that contained the stars.

Pythagoras was a young man when Anaximander was expounding his cylindrical thoughts. Whether or not Pythagoras heard of these ideas I do not know, but if he did hear of them, he would have disagreed: on aesthetic grounds the earth must be a sphere and must follow a circular orbit about the sun. Aristotle, a couple of centuries later, agreed, but on altogether more down-to-earth grounds. Aristotle believed in confirming philosophical hypotheses via observation, and he saw several indications that the planet he resided on was a sphere. The shadow that the earth cast on the moon during a lunar eclipse was circular, and this was true whether the moon was high or low in the sky. The only object which casts a circular shadow when lit from different angles is a sphere. Furthermore, the manner in which a ship at sea disappeared from view, in whatever direction it was seen, argued for a spherical earth. The known stellar constellations appeared lower in the night sky as one traveled farther south—and new constellations appeared in the southern sky.

Eratosthenes (276–195 BCE) is widely known as the founder of geodesy. He is given credit for the first measurement of the earth's radius. For him, qualitative observations were insufficient: he wanted measurements, and this is why he is so highly regarded by geodesists. I will explain his method and that of two later luminaries. Eratosthenes was librarian at the famous library of Alexandria, on the coast of northern Egypt (fig. 2.3). Farther south, on the Tropic of Cancer, was the city of Syene (modern Aswan). Eratosthenes describes a deep well at Syene and tells us that the bottom of the well was lit by sunlight at noon on the day of the summer solstice, and so the sun was directly overhead Syene at that time. In Alexandria, he set up a *gnomon* (a staff or rod set perpendicular to the ground to cast a shadow) and measured the length of its shadow at noon on the summer solstice. He states that Syene was a distance of 5,000 stadia due south of Alexandria. From this information and the length of the rod's

(a)

(b)

FIGURE 2.3. (a) The modern library at Alexandria, Egypt. (b) An old picture of the ancient library, known to Eratosthenes, Ptolemy, Euclid, Archimedes, and many other luminaries from classical antiquity. The original library burned down, probably in the year 48 BCE at the hands of Julius Caesar's legions, and was rebuilt, at least in part, before being destroyed at the end of the fourth century. The modern library is a World Heritage Site. (a) Photo courtesy of Carsten Whimster; (b) image from Wikipedia.

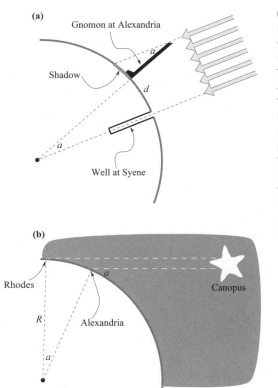

(a)

Gnomon at Alexandria

Shadow

a

d

a

Well at Syene

(b)

Rhodes

a

Canopus

R Alexandria

a

FIGURE 2.4. Two ways of measuring the size of the earth, from ancient Greece. (a) Eratosthenes measured the length of a shadow at Alexandria while the sun was directly overhead at Syene. (b) Posidonius measured the height of the star Canopus (which sat on the horizon when seen from Rhodes) at Alexandria and then calculated the earth's radius using the known distance between Alexandria and Rhodes.

shadow, he derived mathematically the radius of the earth.[4] Eratosthenes' measurement geometry is illustrated in figure 2.4a.

Eratosthenes found the radius of the earth to be 40,000 stadia. The definition of a stadion is ambiguous. It is supposed to be the length of an athletics stadium, but these varied in different parts of the ancient Greek world. The consensus of historians is that a stadion was 185 m, which makes Eratosthenes' estimate for the earth's radius a little shy of 7,400 km. This is too big by 16%, we now know—but it is a great achievement given the limited resources he had available. The main point is that he used measurement and observation in order to quantify; he did not settle for a mythological or a philosophical and qualitative explanation for the earth's physical shape and size.

4. The ratio of the length of the gnomon shadow to gnomon height is tan a. The angle a is also equal to d/R, where R is the radius of the earth (see fig. 2.4a). When a and d are known, R can be determined.

Eratosthenes assumed that the sun was so far away that a ray of sunlight striking the bottom of the well at Syene was effectively parallel to a ray of sunlight grazing the top of a gnomon at Alexandria. This is true. However, he did make a number of errors, which account for the 16% error in his measurement. Syene is not due south of Alexandria (it is 2° 58′ east); it is not quite on the Tropic of Cancer (it is north by 22′). The sun has a finite angular width—it is not a point source of light—and so the length of the gnomon shadow is difficult to estimate accurately. In fact, Eratosthenes measured the angle a of figure 2.4a as being 1/50 of a full circle, or 7° 12′, whereas it is actually 7° 8′. Distances between cities were more difficult than angles to estimate in those days, and his distance of 5,000 stadia between Syene and Alexandria is 10% too large.[5]

A hundred and fifty years later, Posidonius of Apamea was aware of Eratosthenes' work and produced a different estimate of the radius of the earth. On the Mediterranean island of Rhodes, where he lived, he noted that the star Canopus (the second brightest in the sky) just grazed the horizon, whereas farther south in Alexandria, it rose 1/48 of a circle, or 7° 30′. Knowing the distance between Rhodes and Alexandria, Posidonius could then estimate the earth's radius using the geometry shown in figure 2.4b. His estimate was 11% too large, though possibly so close to the true value only because of a fortuitous cancellation of errors. The method he used was reasonable, except that it did not take into account the atmospheric refraction of starlight.

Fast-forwarding a millennium and moving east, we find the caliph Al-Mamun, seventh sovereign of the Abbasids, in Baghdad, sending surveyors north and south of his city, across the plain of Sinjar. Al-Mamun was greatly interested in astronomy, geodesy, and medicine; and Baghdad reached its scientific zenith under his reign (786–832 CE). His surveyors traveled until they reached locations where the angle to the Pole Star was one degree different from the angle observed at Baghdad. The geometry is shown in figure 2.5a. The distances from Baghdad to the two sites were measured with either wooden rods or knotted ropes. From the geometry, the radius of the earth was determined to be 6,364 km (or possibly 6,409 km; the

5. That is, 5,000 stadia corresponds to 925 km if we are right in assuming 1 stadion = 185 m. The distance between Alexandria and Aswan is 842 km, as I found after a two-minute interrogation of Google Earth. Here is an eloquent illustration of the vast increase in geodetic information; how much effort would it have taken to ascertain the distance between these two cities 20 years ago—or 200 years ago?

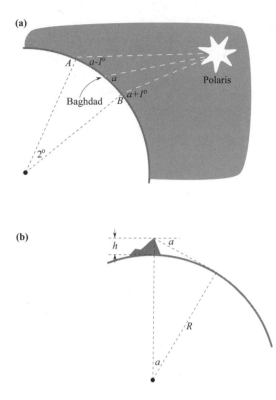

(a)

A $a-1°$
Baghdad a B $a+1°$
$2°$

Polaris

(b)

h a

R

a

FIGURE 2.5. Two early Islamic measurements of the size of the earth. (a) Al-Mamun sent surveyors to A and B, from which the height of Polaris differed by 2°, and measured the distance from A to B. (b) Al-Biruni required only one observer, at the top of a mountain. By measuring the dip angle *a* to the horizon, and knowing the height, *h*, of the mountain, he determined the earth's radius, *R*. (From the geometry we see that $R \approx 2h/\tan^2 a$).

sources are not consistent). The actual mean value of earth's radius is 6,371 km, so Al-Mamun's estimate was extremely good.[6]

Light from Afar

After the Greeks came the Romans. Rome advanced Western civilization in many ways, but science was not one of them. And the history of science in general, and of geodesy in particular, got worse before it got better. The Dark Ages followed the collapse of the Roman Empire, and much of the intellectual life of Europe was put on hold. Even the works of earlier, more enlightened times were forgotten or swept away. From the time of the Greeks until the medieval period, little advance was made in European geodesy.[7]

6. Much has been written about these three (and other) measurements of the Earth's radius. See for example Buttiner and Wallin (1999, p. 97), Evans (1998, chap. 1), Fischer (1975), Harley and Woodward (1987, chap. 8), and Lindberg (1978. chap. 1).

7. An indication of how far European science fell with the eclipse of ancient Greece comes from another of their philosophers, Aristarchus, who lived in the third century BCE.

The one figure who stands out in this period is Claudius Ptolemy, a Roman who wrote in Greek and lived in Alexandria, the home of Eratosthenes 350 years earlier. We will meet Ptolemy again, in more auspicious circumstances. He has been called the father of modern geography because of his contributions to mapping, but his influence on geodesy was baleful. Recalculating Posidonius's results, Ptolemy used a value for the stadion that was only $5/7$ of the what we now believe it to be, and so determined a measurement for the radius of the earth that was 17% too small. He also wrongly believed that Asia stretched halfway around the world instead of a third of the way. The result of these errors came home to roost one and a half millennia later, when Columbus consulted Ptolemy's writings and concluded that Asia was far closer than was actually the case.[8]

During the European Dark Ages, the flame of geodesy was carried forward elsewhere. In Tang China around 725 CE, a government-appointed Buddhist monk named I-Hsing (Yi Xing) set about determining the size of the earth by setting up stations across China to measure the length that corresponds to a degree of arc. Such measurements anticipated Al-Mamun by a century and, we will soon see, became very fashionable in Europe a thousand years later. I-Hsing arranged for measurements to be made of the lengths of the midsummer and midwinter shadows from gnomons and of the altitude of Polaris. From these he made estimates of the length of a meridian arc.[9] This investigation took place in a flat-earth culture and stands out more for the advanced nature of the investigations than for the influence they had on geodesy.

Now we come to Al-Biruni (973–1048 CE, shown in fig. 2.6). His method for estimating the earth's radius was different from the three con-

---

He advocated the view that earth moved around the sun. He also estimated the distance to the moon, as follows. Assuming that the moon went around the earth at constant speed in a circular orbit, he measured the time it took for the moon to pass through the earth's shadow during a lunar eclipse. Knowing the size of the earth and the moon's orbital period, he determined that the earth-moon distance equals 60 earth diameters—pretty close to the true value. Aristarchus's views, and the questions he asked, would not be seriously entertained in Europe again for close onto 1,900 years. To learn more about Aristarchus, see, for example, Heath (1981).

8. You might well ask why medieval Europeans consulted ancient texts to determine the size of the earth, rather than simply go out and repeat the experiments themselves. Good question. Christian dogma may not have been so pervasive as to insist on a flat earth, as we have seen, but it did discourage innovation and independent query. See, e.g., Boorstin (1983).

9. An analysis of I-Hsing's achievements is provided in Beer et al. (1961).

FIGURE 2.6. Abū Rayhān Muhammad ibn Ahmad Bīrūnī (Al-Biruni). Adapted from an old Iranian painting.

sidered so far because it required only one observer at one location, as you can see from figure 2.5b. He measured the dip angle to the horizon from the top of a mountain. Knowing the height of the mountain, he calculated the earth's radius.

The light of science in general, and of geodesy in particular, may have gone out in Europe, but it burned brightly in the Islamic world. Much of the earlier learning of the Greeks was known to Islamic scholars (both Al-Mamun and Al-Biruni read Aristotle, for example) and would return to European knowledge through them.[10]

Battle of the Bulge

The modern era of geodesy began early in the seventeenth century with a Dutch astronomer and mathematician, Willebrord Snell.[11] Snell took advantage of recent developments in instruments and mathematics to develop the system of measuring distances via *triangulation*. The telescope, the theodolite and its ancestors, and the invention of logarithms combined to permit distances over land to be measured with an accuracy unknown in

10. For Al-Mamun and Al-Biruni, see Buttimer and Wallin (1999); *Encyclopedia Britannica*, s.v. "Biruni"; Glick, Livesey, and Wallis (2005, pp. 88–90); Harley and Woodward (1987, p. 141); and Lindberg (1978, chap. 1).

11. Snell is known to physicists in the English-speaking world for his law of optics.

Al-Biruni

The eleventh-century polymath genius Al-Biruni, unknown in the West until modern times, was a true scientist in the modern sense: he made careful observations and measurements and drew conclusions based on these, independently of any prevailing mythology. An Uzbek who lived much of his life in (and died in) Afghanistan, then part of the Persian world, Al-Biruni wrote in Persian and Arabic and spoke several other languages, including Hebrew and perhaps Greek.

Al-Biruni wrote treatises on the astrolabe and other navigational instruments and made a study of map projections. He estimated the distance between the earth and the sun, believed that the earth was a sphere that rotated about its axis, and considered it just as likely that the earth orbited the sun as the other way around. Half a millennium before triangulation became well-known in Europe, he employed it to estimate the earth's radius to within 17 km of the true mean value (such accuracy was not achieved in the West until the sixteenth century). He also measured the rate at which the earth's rotation was slowing down.

Al-Biruni's scientific interests extended beyond geodesy to astronomy, mathematics, mineralogy, and optics. Outside of science, he was keenly interested in Islamic law, India, linguistics, and astrology, the last of which he refuted repeatedly.

Europe since classical antiquity. I will have much more to say about the instruments of surveying and about triangulation in a later chapter. Here, I introduce triangulation as a finished product, a tool that came to be used throughout the seventeenth century to make increasingly accurate geodetic measurements. Indeed, Snell himself used triangulation to survey the distance between two Dutch towns in 1615 and from his data was able to estimate the earth's radius to within a few percent.

In 1669 a French priest, Jean-Felix Picard, used triangulation to form an accurate map of the area around Paris. This was the beginning of a monumental survey that would map the whole of France—the first survey to cover an entire country. Picard also wanted to measure the length of a meridian arc (a line stretching north-south that covers at least a degree of longitude) and so estimate the size of the earth, using Snell's methods but improving on his result. Picard chose as his southern point Malvoisine, an estate which lay some 30 km south of Paris (today, an obelisk in Malvoisine

FIGURE 2.7. Looking at stars to find out the shape of the earth: The Cassini Room at the Paris Observatory, with the Paris meridian traced along the floor. Inset: Giovanni Cassini, director from 1671 to 1712. From Wikipedia, courtesy of FredA.

marks the exact spot). The northern point for his survey was Sourdon, about 110 km due north (another obelisk). The latitudes of these end points were determined by astronomical observations, and the distance between them was measured by triangulation. The result was an estimate of the earth's radius of 6,329 km, which is less than half of one percent below the currently accepted mean polar radius.

At this point, the Cassinis enter our story. Giovanni Cassini was an Italian astronomer who moved to Paris at the time of Picard's survey to take over the newly formed Paris Observatory (fig. 2.7). Over the next century, four generations of Cassinis would dominate the observatory and also the Académie Royale des Sciences. They would eventually produce an accurate map that showed France to be significantly smaller than had been thought. Over the period 1683–1716 Giovanni Cassini and his son Jacques extended Picard's meridian arc southward to Collioure, while Philippe de Lahire extended it northward to Dunkirk. These measurements resulted in two estimates of the length of a degree longitude (that is, of the length of 1° of arc running north-south). The length of a degree north of Paris was found to be less than the length south of Paris (see the caption to fig. 2.8a for the reason). Because they felt that the survey had been conducted very accurately, the Cassinis concluded that the earth was not a sphere, but was instead a prolate spheroid, stretched at the poles like an egg. This was the first evidence that the earth was not spherical.

We have seen that Newton also thought the earth was not quite spheri-

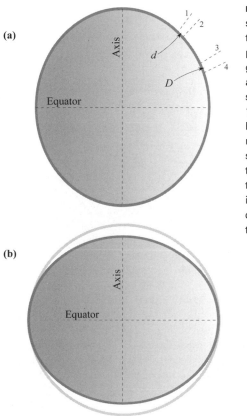

FIGURE 2.8. Prolate and oblate spheroids. (a) The Cassinis believed that the earth was stretched at the poles. (The stretch is greatly exaggerated here.) The dashed lines 1–4 are all perpendicular to the prolate spheroid; if the angle between line 1 and line 2 is the same as the angle between 3 and 4 (say 1°), then the meridian arc length d will be shorter than D, because the curvature of the surface is greater at 1–2 than at 3–4. (b) This oblate spheroid is flattened by $\frac{1}{5}$. The earth is much closer to being a sphere: it is flattened by about $\frac{1}{300}$.

cal, but in the opposite sense: he thought that the earth was squashed at the poles (fig. 2.8b). The eminent Dutch physicist Christiaan Huygens agreed with Newton, and thus, the battle lines were drawn.[12] The Cassinis insisted on an earth stretched at the poles, based on their observations, whereas Newton and Huygens believed in a squashed earth, based on Newton's theory. Who was right?

Doubts about the validity of the Paris data grew within the French camp, and two expeditions were sent across the world in 1735 and 1736 to settle the matter. One expedition, headed by Charles Marie de la Condamine and Pierre Bouguer, traveled to Peru to measure the length of a 1° meridian arc near the equator.[13] A second expedition under Pierre-Louis Moreau de Maupertuis conducted a survey in Lapland to determine the length of a 1° arc near the North Pole. The surveys were conducted with unprecedented accuracy, and the result was unequivocal: the earth was flattened at the poles, as Newton and Huygens (and several French surveyors, including Maupertuis) had thought. The eminent French philosopher Voltaire got involved in the acrimonious aftermath of these expeditions, saying to Maupertuis, "You have flattened the earth and the Cassinis." He said to a member of the other expedition, "You have found by prolonged toil what Newton had found without even leaving his home."[14]

The expedition results may have been painful to the Cassinis, but the important point to note is that science was being conducted properly. Experimental evidence determines the facts; theories are good, but only if supported by observation; bad data are acknowledged and fixed. Egos may influence the discussion (scientists are people, after all) but not the process. Here we have an early example of modern scientific method, right down to the large cost of the expeditions.

12. Cassini and Huygens were both honored in 1997 with the launch of the NASA/European Space Agency Cassini-Huygens satellite mission to Saturn.

13. The site is in modern Ecuador, which did not then exist as an independent nation. Indeed, when Ecuador obtained its independence in 1830, its name was chosen, so it is claimed, in part because of the prestige bestowed on it by the French geodetic mission to its capital, Quito, almost a century earlier.

14. The quotes are from Smith (1997). To read further on the Picard and Cassini surveys, and on the controversy about the shape of the earth (which displayed a great deal of ego, animosity, petulance, intrigue, ambition, and professional rivalry, as well as extraordinary effort, determination, and immaculate surveying techniques), see Danson (2006), Jardine (1999, chap. 6), Konvitz (1987), Murdin (2008), Terrall (2002), and Whitaker (2004, chap. 7). Long after the French expeditions, it was shown that there were errors in their calculations that exaggerated the flattening effect.

Ironically, the first experimental evidence for an oblate-spheroidal earth, as advocated by Newton and Huygens, came from a Frenchman who was acting under Giovanni Cassini's instructions, 35 years before the expeditions, at the time of Picard's survey. In 1671 Jean Richer had been sent to Cayenne, French Guiana, to measure the parallax of Mars (to estimate its distance from earth). Richer noticed that his pendulum clock, which was accurate in France, ran slow in Cayenne. In fact, it was so slow that the pendulum had to be shortened by 2.8 mm to fix the problem. Newton and Huygens used Richer's pendulum data to show that the earth's gravity was weaker near the equator, thus showing that our planet is oblate, not prolate.

We use the word *ellipsoidal* to describe the shape of the earth. Unlike *spheroidal*, it has a precise mathematical meaning, and it is the term of choice today. So, henceforth in this book I will refer to an ellipsoid and not to an oblate spheroid when referring to the shape of the earth. A sphere is defined by a single parameter, the radius, whereas an ellipsoid requires two parameters to define it: its radius and its *eccentricity*. The latter refers to the degree of flattening. A small eccentricity means that the ellipsoid is very nearly a sphere; a large eccentricity means that the sphere is more squashed.

Figure 2.9 shows the result of earth shape measurements made after the seminal French arc length measurements of the 1730s. Note how the increasingly accurate experiments (due to increasingly sophisticated measuring equipment, as we will see) converge over the decades to the same values. From the 1930s, the measured flattening of the earth has arrived at a consensus value of about one part in 298; from about 1850 the measured radius of the earth has converged on a value that is within a couple of hundred meters of 6,377.2 km.

Vertical Deflection

The perceived shape of the earth had changed over the centuries from flat, to spherical, to ellipsoidal, as surveyors ranged farther and farther afield and employed better and better equipment. The methods used for geodetic measurements in the seventeenth century, still used in the twentieth century, involved celestial observations and triangulation. (I will discuss the tools and methods in detail over the next couple of chapters.) Today, we know rather precisely that the best ellipsoid to fit the shape of the entire planet has a flattening of one part in 298.257223563. Yet, even by the early

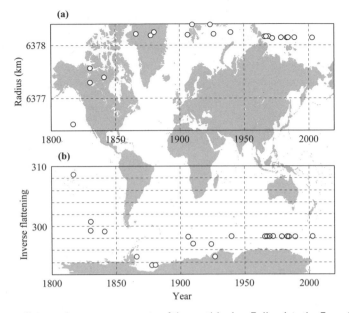

FIGURE 2.9. Converging measurements of the earth's size. Following the French meridian expeditions of the 1730s—which found the earth's radius to be 6,397 km and an inverse flattening of 191—there were many other measurements of the earth's radius and flatness. Note the agreement of recent measurements. (The flattening factor is simply the fractional difference between polar and equatorial diameters; inverse flattening is the reciprocal of this fraction.)

nineteenth century, surveyors were beginning to realize that a smooth, simple ellipsoid (simple because only two parameters are required to define it) was not going to be good enough for very accurate geodetic work. Bouguer, on the 1736 Peruvian expedition, noted that a plumb line was pulled in the direction of nearby mountains (the Andes). Later, we will see that another great survey found the same kind of deflection. Out in the field, surveyors were beginning to see the effects of local concentrations of mass.

Back home, mathematicians recognized from the early nineteenth century that an ellipsoidal model of the earth was no longer tenable: contradictions within the data they analyzed greatly exceeded the observational accuracy of the surveyors' instruments. In 1828 Carl Friedrich Gauss first described the geoid, which is, as we saw in chapter 1, the surface that is perpendicular to all plumb lines. The deviations from the vertical that surveyors were beginning to detect in the eighteenth and nineteenth centuries demanded such an object: local anomalies simply could not be accounted for by the smooth and simple ellipsoid. Gauss wrote: "What we call the surface of the Earth in a mathematical sense, is nothing else but that surface, which everywhere intersects the direction of gravity at right angles, and of which the surface of the ocean is a part. The direction of gravity at every point is determined by the shape of the rigid part of the Earth and its uneven density."[15]

What Gauss called his "figure of the Earth" was clearly the geoid, though the term and idea would not become standard until later in the nineteenth century. Friedrich Robert Helmert, writing in 1880 and 1884, laid the foundations of modern geodesy on the framework of the geoid, a term he coined. Today we know that the geoid is more than Gauss intended because we can measure the effects of lunar gravity and can see that the earth's surface is continually moving, so we define the geoid as an average surface. It includes local gravity anomalies—a mountain range to the north, for instance—but also other effects that are not necessarily understood. For example, there has been an extended discussion in the technical literature about whether an ellipsoid is the best theoretical shape—local anomalies apart—to describe the earth. Perhaps the equator is not exactly a circle; perhaps it is an ellipse. Perhaps the curvature at the South Pole is less than the curvature at the North Pole, giving the earth a slight pear shape.

These effects, if they exist, are due to the complex gravitational interaction of the earth with all the other bodies in the solar system. Because they are not well established in the minds of theorists, these tiny gravitational effects have not yet been applied to modify the reference ellipsoid, but have been added to the geoid. To this extent, the geoid is a catchall for everything we don't understand about the earth's shape. The difference between the ellipsoid and the geoid is the difference between what we understand about the shape of the earth and what we measure. That the difference is small—less than a couple of hundred meters over the entire surface of our planet—tells us that geodesists have a pretty good idea about why our earth is the way it is.[16]

Over a four-year period from 1821, Gauss conducted a detailed survey

15. Howarth (2007). Gauss was a mathematical genius of the first magnitude who concerned himself with many of the applications of his math, including geodesy and surveying. We will encounter him again in this book.

16. For the history of modern geodesy, see, e.g., Hoare (2005), Smith (1997), Torge (2001), or Wilford (2000). In addition, the USGS and NOAA maintain educational websites on the subject.

of the newly established kingdom of Hanover in northern Germany. He employed triangulation (using 26 triangles) and invented the *method of least squares* to minimize the observation errors. Use of the least squares method—of which, more anon—quickly became standard surveying practice. It provides the best-fit solution to a system of equations with more measurements than unknowns—a situation that applies to surveying and to many other fields that involve observation or experiment. Gauss's interests were so broad, however, that a rumor arose concerning his motivation for the survey. Mathematicians had just discovered the possibility of other types of space, weird curved spaces that later would be taken up by Einstein. In fact this is just a myth: Gauss knew of the possibility of such non-Euclidean spaces but never doubted that the space he surveyed was that of Euclid.[17] The contributions of Gauss and others (such as Euler, Lagrange, and Fourier) during the eighteenth and nineteenth centuries provided us with most of the mathematical tools that are used in geodesy today.

Geodesy in the Electronic Age

Modern technology has changed the way that we do geodesy in the twenty-first century. Indeed, modern technology has led to a redefinition of the scope and applications of geodesy. Lasers are used to estimate the distance between A and B with very high accuracy, whether these two points define a residential lot boundary or the nearest points of the earth and the moon (fig. 2.10). Radar altimeters and, more recently, satellite-borne interferometric radars allow us to detect distances, and the slow change of distances with time, to within a fraction of a centimeter, as we saw in figure 1.11. Satellites and computers greatly assist our employment of these high-tech measuring sticks, so that today we can measure and remeasure the entire surface of the earth every few minutes, every day, continuously. Such unprecedented monitoring capability—it is like a giant stethoscope measuring the breathing and the heartbeat of an enormous beast—has shown us that the surface of the earth is not stationary. It moves cyclically (due to

17. Euclidean space is the space we all learned about in school, in which the angles of triangles add up to 180°. In non-Euclidean spaces, the angle can add up to more than 180° (think of a triangle drawn on the surface of a sphere) or less (a triangle drawn on a saddle). Gauss's geodesic triangle added up to 180° with an error of less than two-thirds of a second of arc. See Breitenberger (1984). For more on the history of the least-squares method, see Chambert (1999, chap. 9) and Sheynin (2004).

FIGURE 2.10. An LR-3 laser ranging retro-reflector on the moon. The retro-reflector is the white object on the surface between astronaut Buzz Aldrin and the lunar module *Eagle*. Photo by Neil Armstrong, July 20, 1969, courtesy of NASA.

lunar gravity, for example) and also drifts slowly and quickly (tectonic plate shifts, for example).[18]

The first useful radar (Chain Home, deployed in England just before the Battle of Britain) appeared in 1940 and the first useful satellite (Sputnik), in 1957. Lasers and computers emerged during the last half of the twentieth century. It is the combinations of these devices that have brought geodesy to its present high state; the best-known modern high-tech application is the GPS system of satellites. Because the new geodesy has shown us that the shape of our earth changes, geodesists have had to redefine and refine their measurement of the passage of time. There are

18. A recent study by NASA and the University of Texas has shown that a lot of the changes in the earth's surface are due to changes in climate, particularly during El Niño years, when large bodies of water get shifted about. See, e.g., Adam (2002) and Leuliette et al. (2002).

Earth Wobbles

There are two types of earth wobbles—by which I mean shifts in the orientation of the earth's rotation axis. The first is due to the tidal forces of the moon and the sun. This wobble moves the orientation of the earth's rotation axis relative to the stars and consists of several components. The largest and slowest component is the precession of the equinoxes, which has an angular amplitude of 23½° and a period of about 26,000 years, as we have seen.

The second type of wobble is quite separate—it would occur even without lunar and solar tidal forces. The largest component in this second type of wobble is a consequence of the fact that the earth is not quite spherical. Physicists call this wobble *free nutation*; geodesists refer to it as *polar motion* or as *variation of latitude* because it moves the lines of latitude. Happily, this movement is small and slow. Polar motion is known to consist of three

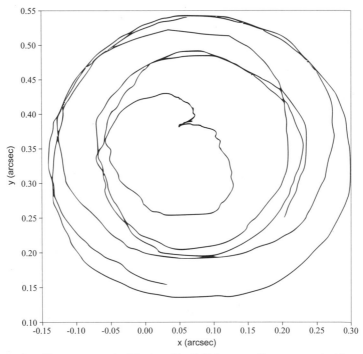

Polar motion. The movement of the true North Pole over a five-year period from January 1, 2004. The coordinate system aligns the *x*-axis east-west, such that movement to the right is toward Greenwich. One arc second corresponds to a distance of about 30 m. Data from the Earth Orientation Center, International Earth Rotation Service.

contributions: one is a slow drift and two are circular periodic motions, of different periods.

The North Pole has drifted about 20 m over the last century; the cause is not well understood. The periodic motions consist of an annual cycle and a larger *Chandler wobble* with a period of about 433 days (the period is not quite constant). Isaac Newton predicted the Chandler wobble, and later in the 1700s Leonhard Euler worked out the theoretical details, assuming a solid earth. Euler predicted a period of 305 days. Later, when the wobble was measured experimentally by American astronomer Seth Chandler, the longer period was a puzzle until Simon Newcomb (a Canadian-American astronomer) repeated the calculations in 1891 while allowing for the fluid nature of the earth's core. This explained the discrepancy.*

Polar motion has been measured with exquisite accuracy since 1962. In the figure you can see a five-year section of the data. The largest component of this motion is Chandler wobble, with a radius of about 0.3″ (corresponding to a diameter of about 9 m).

In addition to lunar and solar tidal wobbles and free nutations, there are daily and twice-daily wobbles due to the ocean tides. This contribution to the earth's wobble was unearthed only within the last couple of decades. These movements are measurable despite being very small—less than 3 cm.

* The source of power driving the Chandler wobble was discovered only in 2001: the wobble is perpetuated by fluctuating pressure in the ocean bed and atmosphere.

---

four basic ways that time can be measured: by reference to the earth's rotation, by reference to the orbits of earth and of other planets and satellites within the solar system, by reference to pulsars, and finally—rejecting celestial and planetary observations—by reference to atomic clocks.

The first of these is the oldest: for millennia, farmers and priests have watched the sun rise and move across the sky as the earth turns. Greenwich Mean Time (GMT) was based on the earth's rotation. But we now know that the rotation rate changes with time: we saw in chapter 1 that it drifts over centuries and oscillates minutely over an interval of a few hours. GMT has now been discontinued; its successor is Universal Time (UT1), which incorporates some of the known rotation modifiers, such as polar motion (see "Earth Wobbles"). Even so, the rate at which a UT1 clock ticks is not constant. The second method of measuring time, by observing the orbital motion of other members of the solar system, was employed for a couple of centuries after the invention of telescopes to determine longitude, as we will see. (The phases of the observable moons of Jupiter were used to synchronize astronomical observations taken at different points on the earth.)

The third and fourth methods are entirely modern. Certain atoms vibrate at a fixed frequency and so can be harnessed to provide timekeeping that is independent of the outside world. (Well, not quite independent, Einstein would remind us.) Pulsars are fast-spinning neutron stars that emit beams of very-high-energy electromagnetic radiation. They pulse at a very regular rate, so regular that they rival atomic clocks.

Thus, time and position can be measured very accurately, enabling us to closely monitor distances, speeds, and rotation rates. Geodesy has expanded its applications as a result of this increased capability. The old aims still apply: geodesy still deals with the measurement and representation of planet Earth and its gravitational field, and it still is used for mapping. However, geodesy is today applied to other fields such as ecology (the movement of ground due to melting ice caps, mining, or waste disposal); engineering (e.g., for the planning of dams and the headwaters they create); urban management; geophysics; and hydrography. As accurate monitoring and measuring of our planet continues and improves, the applications of geodesy are likely to expand.

CHAPTER THREE

Surveying

We have seen the results of geodetic surveys. Now we get down to basics and see how surveys are done on the ground (or the oceans) by examining the tools of the trade and describing how these tools and the science of surveying evolved from peg and rope to GPS.

The First Surveying Instruments

Surveying has accompanied activities such as farming and empire-building since the dawn of history. Farmers needed to know where their land joined a neighbor's; conquerors needed to inventory the assets that they had acquired. Consequently, surveying tools have been around for a very long time. Later and more sophisticated tools are developments of earlier, simpler ones, as you will see. Here, I survey the development of surveying tools.[1]

PEG AND ROPE

The ancient Egyptians of about 3000 BCE used ropes attached to pegs to lay out land. Recent research suggests that Stonehenge (constructed around 2500 BCE) was also laid out with peg and rope technology. Given the cultural and geographical distance between the ancient Egyptians and the mysterious people who built Stonehenge, it seems likely that the peg and rope idea was widespread in Europe by the third millennium BCE. Applying a length of rope attached to a peg suggests strongly an appreciation of plane geometry because it is equivalent to the compass-and-ruler type of geometrical construction that I (and millions of others who exceed

1. For early surveying instruments, see, e.g., Lewis (2001).

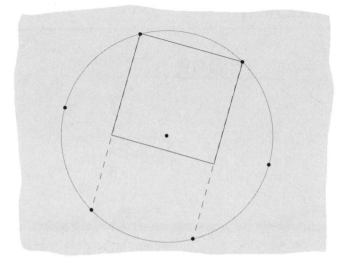

FIGURE 3.1. Constructing a square on level ground, using only a peg and a rope. From the central peg, mark out a circle. Starting at a point on the circle, stretch the rope to another point on the circle; repeat until the circle has been divided into six equal segments. The dashed lines are parallel. Stretch the rope along these lines from the circle to construct the square shown.

a certain age) was taught in grade school. Consider, for example, figure 3.1, in which you can see how a peg and rope system can be used to construct a square. This method is just one way of doing it; there are many others.

The method of figure 3.1 is simple in that it requires no knowledge of algebra. Thus, it does not assume that the surveyor is aware of Pythagoras's theorem. But with such extra knowledge, the construction of a square would be even simpler: the surveyor would equip himself with two ropes, one the length of a side of the square he sought to lay out, and the second with a length greater by a factor of $\sqrt{2}$. Two such ropes make the construction of a square quite trivial. It is likely that the mathematical knowledge of third-millennium-BCE Egyptians was more sophisticated than that of their near contemporaries in stone age Britain,[2] and yet we see from figure 3.1 that even an intuitive knowledge of basic plane geometry, unsupported by theoretical ruminations, is sufficient for constructing squares.

2. For example, the Egyptians knew about 3-4-5 triangles—that is, if the lengths of the sides of a triangle are in the ratio 3:4:5, then one of the angles of the triangle is 90°. That they used geometry when surveying is pretty obvious from their pyramids, which have very square bases. As for the builders of Stonehenge, the theory that they surveyed the construction site using peg and rope techniques is due to Johnson (2008).

Later, the peg and rope combination used for land surveying would be replaced by the chain and compass. A *Gunter's chain*, developed in early-seventeenth-century England (fig. 3.2), consisted of 100 iron links and was 66 feet long. This odd unit of distance survives today as the length of a standard cricket pitch. A compass was used to determine direction, with an accuracy of one-quarter of a degree at best (relative to magnetic north, of course). Between them, the chain and compass enabled land surveyors to estimate distance and direction.

GROMA

The Romans were not great scientists, but they excelled at construction engineering, and this required surveying. The basic tools of Roman surveyors were the water level and the *groma*, illustrated in figure 3.3. Very likely the groma originated with the Babylonians, long before the Romans. It was used to establish lines and perpendiculars to those lines, as when

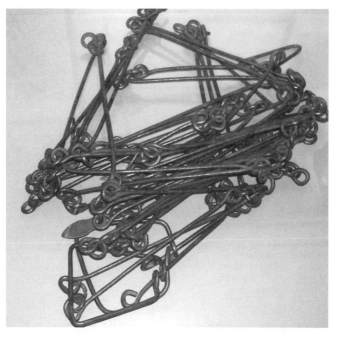

FIGURE 3.2. A Gunter's chain. Named for the clergyman and mathematician who invented it around 1620, this standard surveyor's chain was 66 feet long and consisted of 100 links. A chain is the length of four *poles*. Ten chains make a *furlong*, with eight furlongs per mile. An acre is thus one chain by one furlong. From Wikipedia, courtesy of Roseohioresident.

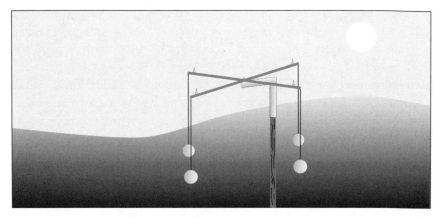

FIGURE 3.3. A groma. A vertical rod holds two perpendicular horizontal rods, offset as shown. From the end of these are suspended four plumb bobs. The main rod is set vertically in the ground, enabling the surveyor to sight along horizontal lines at right angles.

laying out a military camp or a building. One stunning example of Roman construction that survives to the present day is the Pont du Gard aqueduct in France, a World Heritage site that was constructed around the year 19 BCE. The distance from Roman Ucetia to Nemausus (modern Uzès to Nîmes) was 50 km, and the drop in elevation of the aqueduct was only 17 m, corresponding to a mean gradient of ⅓₀₀₀. Thus, the Romans must have accurately measured differences in elevation to within a minute of arc. An aqueduct is unforgiving of errors in layout or construction: water will not flow uphill, should the surveyors make a mistake or the foundations settle. Clearly, a great deal of accurate surveying must have been performed prior to constructing this aqueduct, yet the Romans used only the groma and water level. Several such aqueducts exist across Europe (fig. 3.4); they are as much a testament to Roman surveying skills as to their engineering prowess.[3]

ALIDADE

An *alidade* is an instrument so basic that you may be surprised to learn that it has a name. At its simplest, an alidade is a straight rod such as a ruler, with a raised sight at each end—like a rifle gunsight. The ancient Greeks wrote about it in the third century BCE, calling it a *dioptra*. Our name

3. The Pont du Gard aqueduct has for centuries been held up as an icon of precision engineering. See Holmes (2010, pp. 354–55).

(a)

(b)

(c)

FIGURE 3.4. Roman aqueducts through the ages. (a) A 1786 painting of the Pont du Gard aqueduct, in France. (b) An old print of the Segovia aqueduct in Spain. (c) A recent photo of the Valens aqueduct in Istanbul. All three structures still exist.

(a) *Le Pont du Gard*, by Hubert Robert; (b) *Encyclopedia Britannica*, 1911 ed.; (c) Wikipedia photo by Necdet Cevahir.

FIGURE 3.5. A modern alidade, used for estimating the direction of the object viewed. U.S. Navy photo by Mass Communication Specialist Seaman Jerine Lee.

comes from the Arabic word for "ruler." Surveyors often used it in conjunction with a *plane table*, a flat horizontal surface on which paper can be placed, to draw a plan of the surrounding area. After placing the alidade on such a table, the surveyor sighted along the instrument toward a nearby feature of interest, be it a building, a tree, or a peg driven into the ground; he then drew a line on the paper guided by the alidade ruler. In this way many features could be placed on the paper in their correct positions. We still use alidades today—for a modern one, see figure 3.5—though the technology has changed considerably. The instrument is usually paired with an angular scale so that the direction of a feature of interest can be measured.

ASTROLABE

Astrolabes are ancient machines for calculating celestial movements. They have been around since classical antiquity and were popular in the east during the Islamic Golden Age, say from the eighth to thirteenth centuries. In Europe they were widely used until the Renaissance. There were several variants, of which the *planispheric astrolabe* was the most common (fig. 3.6). This instrument was applied to solve various astronomical prob-

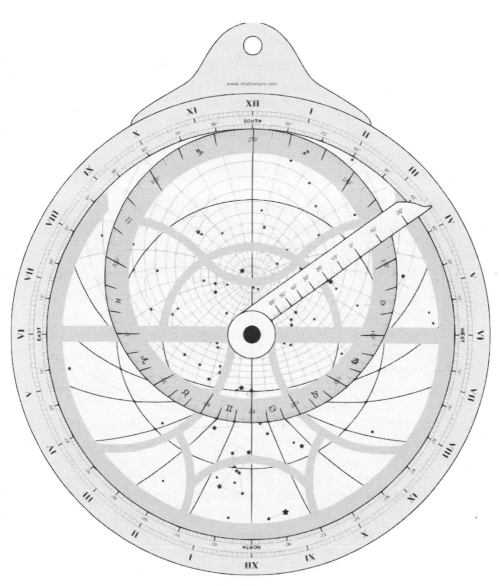

FIGURE 3.6. A planispheric astrolabe, used for predicting the position of celestial bodies at a given time, or, by observing the sky and comparing star positions, for telling the time. Adapted from a Wikipedia image.

lems without calculation, in the centuries before the telescope was invented and before planetary motion was properly understood, by presenting a view of the relative positions of the sun, the moon, the planets, and the stars at different times. That is, the astrolabe showed what the sky looked like at different times and places. Astrolabes were usually made of brass and were accurate only for a particular latitude. They were used to tell the time of day, to determine when sunrise or sunset would occur, or to point the faithful toward the direction of Mecca.[4]

Astrolabes were astronomical instruments, not really navigational or surveying tools, and I mention them here only because of their antiquity and because they gave rise to the mariner's astrolabe during the period leading up to the European Age of Exploration, to which we now turn.

Sunny Latitudes

Many of the navigational instruments that were derived during the first 200 years of European maritime exploration of the world[5] were designed to estimate latitude from the height of the midday sun. We will examine these instruments in this section. Mapmakers as well as navigators used them, and they led to the development of more sophisticated instruments employed in later times for land surveying as well as maritime navigation.

CROSS-STAFF

Also known as a *Jacob's staff*, the simplest form of the cross-staff—which I describe here—was used by mariners to estimate the height of the sun, from which they could determine latitude. Land surveyors made use of a more sophisticated version.

The cross-staff was first described in the fourteenth century, though the instrument may be much older. It consisted of two straight rods set perpendicular to each other. The larger *main staff* supported the smaller *transom*, which could slide along it, as suggested in figure 3.7. The angle subtended by the ends of the transom at one end of the main staff, where the mariner placed his eye, was marked on the main staff. A mariner who wanted to estimate the height of the sun would align the lower end of the transom with the horizon, as shown, and the upper end with the sun.

A simple instrument, the cross-staff was accurate to within a degree or

4. The astrolabe is discussed by Morrison (2007).
5. Roughly the fifteenth and sixteenth centuries and known as either the Age of Exploration or the Age of Discovery.

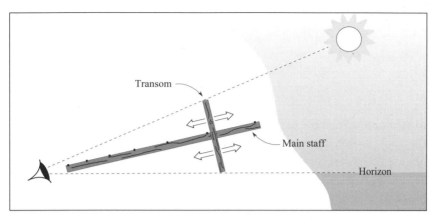

FIGURE 3.7. A cross-staff. The user lines up the sun along the upper edge of the transom and the horizon along the lower edge. The sun's height is read off a scale on the main staff. The cross-staff can also be used, perhaps more comfortably, to measure the angular distance between two stars.

two, but was difficult to use.[6] Not least of the problems was the necessity of looking at the sun directly. Also, a mariner would have to hold the staff steady so that it was aligned with the horizon while he switched his gaze to the sun. Quite apart from this difficulty, the eye's movement introduced an error, especially for large angles. (In practice, this limited the angle being measured to less than 50°.) One advantage of the cross-staff compared with some later instruments was that it was independent of gravity—it did not require a plumb bob—and therefore swayed less on a pitching deck. In other words, the angle of the sun was measured relative to the horizon (not to the vertical), and both sun and horizon pitched in the same way.

QUADRANT

From the fifteenth century, the quadrant consisted of an alidade attached to a quarter circle (hence the name) graduated in degrees, and a plumb bob. The instrument was used to estimate the sun's noonday altitude, both for navigation and early mapmaking. You can see how it works from figure 3.8. Clearly, like the cross-staff, it would be difficult to use for estimating latitude because the user would be obliged to look at the sun. The plumb line was an improvement on land because it provided an absolute

6. Errors in measuring the height of the sun with a cross-staff included instrument parallax, the angular width of the sun, and horizon dip (due to atmospheric refraction). These errors were addressed by Edward Wright, who published *Certain Errors in Navigation* in 1599.

reference for measuring the sun's height, and on land the horizon might not be visible, thus rendering a cross-staff useless. The plumb line was a disadvantage at sea, however, because it meant that the angle being measured would change as the ship pitched.

MARINER'S ASTROLABE

Much simpler than the planispheric astrolabe, from which it was adapted in the fourteenth century, the mariner's astrolabe consisted of a circle that was graduated in degrees and a movable alidade that rotated about the center (fig. 3.9a). The alidade was aligned with the noon sun so that latitude could be determined. The instrument was typically made of brass, was 7–8 inches (20 cm) in diameter, and had a thumb ring at the top by which it was suspended. It was made heavy, particularly the lower part, so that it would not swing much in a wind. As you might imagine, this instrument was difficult to use at sea, so mariners who were known to use it, such as Francis Drake, may have obtained more accurate sightings while on land. Also, the mariner's astrolabe accuracy increased with size because angles could be marked more accurately on a larger circle. Drake may have suspended a large astrolabe from a tripod while on land to obtain more accurate readings than were possible on deck. Despite its clumsiness, over time the mariner's astrolabe largely replaced the cross-staff and the quadrant.

The angle measured by a mariner's astrolabe, the *instrumental altitude* (angle *a* of fig. 3.9b) was converted to altitude by consulting an astro-

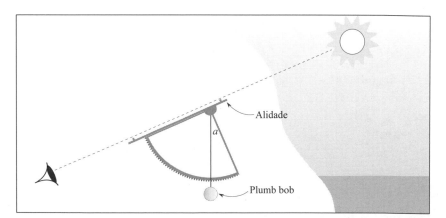

FIGURE 3.8. A quadrant. No reference to the horizon is needed, so this instrument can be used in places where the horizon is not visible. The sun's height in the sky is the angle *a*.

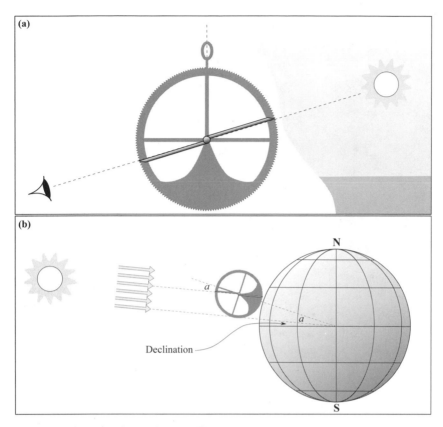

FIGURE 3.9. A mariner's astrolabe. (a) The instrument is suspended to hang vertically, and the rotating alidade is aligned with the sun. Solar height can be measured from a scale marked on the outer circle. (b) Like the cross-staff, the quadrant, and the back-staff, the astrolabe must be used in conjunction with tables so that solar declination can be taken into account.

nomical almanac that included tables of solar declination predicted ahead for several years. The declination angle is shown in figure 3.9b. It varies throughout the year, as we saw in chapter 1, being 23.45° at the solstices and 0° at the equinoxes.[7] Latitude was determined from the solar declina-

7. Drake took with him on his famous voyage (see chap. 6) Bourne's 1574 tables for the years 1573–92. Such a set of tables was called an *ephemeris*. Even in those days, more than a century before Newton, it was possible to predict the positions of astronomical objects. As with the planispheric astrolabe, however, the predictions were based on observations, not on the theory of gravitation. One of the reasons for inaccuracy of latitude estimation in Drake's time, apart from instrument error, was the inaccuracy of ephemeris data. For example, the position of the Pole Star was not accurately determined until the end of the sixteenth century.

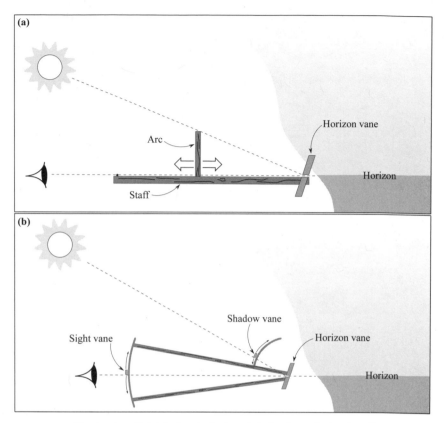

FIGURE 3.10. The backstaff. (a) A simple design showing how the backstaff permits the user to measure solar height without having to look directly at the sun. (b) The Davis quadrant, a variant that was easy to use.

tion and the measured instrument altitude. Navigators who used any of the instruments available during this period (not just the astrolabe) needed to account for solar declination.

BACKSTAFF

Backstaff is the term applied to a family of instruments that were in use for a long period. Among these was the *Davis quadrant*, or *English quadrant*, which replaced the astrolabe as the instrument of choice for measuring latitude at sea. The simplest backstaff, which conveys the basic idea, is shown in figure 3.10a. The *staff* (an alidade) supported a moveable *arc* (often curved) that slid along it. A *horizon vane* showed the shadow cast by the arc and had a hole in the center to permit a mariner to see the horizon.

The arc was adjusted so that the top of its shadow just touched the hole in the horizon vane. The staff was graduated to show the height of the sun (in degrees) for the adjusted arc position. A version of the backstaff invented by Captain John Davis in 1594 is shown in figure 3.10b. The two arcs were fixed, with a common center. A moveable vane slid along each arc so that the shadow and the horizon could be lined up.

The main advantage of backstaffs for observing solar height was that the mariner did not have to look directly at the sun. One disadvantage was that the shadow was cast by the upper limb of the sun, not the center, so an error was introduced that had to be compensated for. Also, the backstaff was no good for measuring the altitude of stars.

Early Optical Instruments and Increasing Accuracy

We now move into the eighteenth century and from there to the instruments of yesterday, if not today. You will see how the older instruments, or components from them, were reused or modified for the new measuring tools, and how new technology and new ideas led to both improvements in accuracy and reduction in size.

OCTANT

The octant was another instrument developed to make celestial and horizon measurements so as to estimate latitude. Its basic elements are shown in figure 3.11. The arc covered 45°, or an eighth of a circle (hence the name), but because the instrument used mirrors, the angles were doubled up, which gave rise to an alternative name: the *reflecting quadrant*. The mirrors of early octants were made of polished metal. The octant was developed in the early eighteenth century by at least four people, independently. Most credit is given to John Hadley, an English mathematician, and Thomas Godfrey, an American glazier, who both came up with the idea in 1731.

By this period, optical instruments such as the telescope and microscope were becoming quite refined, with much of the gross distortions of earlier optics removed through improved glassmaking and lens grinding. So, surveying and navigating were ripe for the introduction of a sighting scope, and these first appeared in octants, as shown in the figure. The index mirror and index arm rotated together about an axis that was the center of the graduated arc. The clever aspect of octant design was the second, or horizon, mirror, mounted on the octant frame so that light from the object being measured (a star, in our illustration) is reflected off both mirrors and

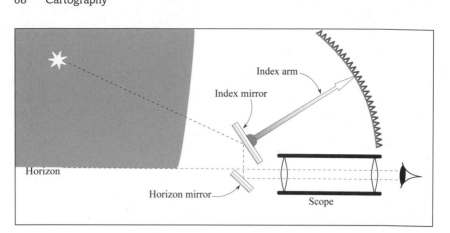

FIGURE 3.11. The octant. Because it used mirrors, thereby doubling angles, the octant covered the same range of angles as the earlier quadrant. It also achieved the same accuracy as earlier instruments that were twice the size. Like the backstaff, the octant measured a relative angle (between horizon and star, here) and so was quite easy to use on board a ship. Unlike the backstaff, it could sight stars as well as the sun.

into the eyepiece of the sighting scope. The navigator or surveyor could see the horizon directly and the star alongside it. This meant that the eye need not be moved during observation: the index mirror angle was simply adjusted until the star and horizon were aligned.

The double mirror made the octant much easier to use than previous instruments for estimating latitude. It ascertained latitude from the height of point-like stars, rather than from the sun, thus reducing a major source of measurement error. High-quality octants were constructed of brass; less expensive, lower-quality instruments were made from wood and ivory. An octant was half the size of a Davis quadrant, but with similar accuracy. When a Vernier scale was added, the octant was accurate to within one minute of arc. The octant and its successor, the sextant, almost eliminated the earlier solar-height instruments by 1780.

VERNIER SCALE AND INSTRUMENT ACCURACY

In the 1630s the French mathematician Pierre Vernier developed a very simple and successful supplementary scale that greatly improved the accuracy of reading that can be made from a measuring instrument. (Vernier's scale was based on earlier work by a Portuguese mathematician, Pedro Nunes, whom we will meet again.) So successful was his idea that it is everywhere today: there are hundreds of examples in your local hardware store.

Consider the calipers of figure 3.12. The upper, fixed-length scale divides each unit into tenths. In the case of calipers without a Vernier scale, there would be only an arrow on the moveable arm to mark the dimension of the hexagonal bolt being measured, and its size would be read off from the upper scale to the nearest 0.1 unit. With a Vernier scale, which is the moveable scale shown on the bottom, the dimension can be read with ten times the accuracy, by aligning the graduation marks on both upper and lower scales. The trick that Vernier exploited is to space the marks of the moveable scale by units of 0.09 instead of 0.10. Thus, if the moveable scale slides a distance of 0.01 unit, the first Vernier mark will align with the first fixed mark because $0.01 + 0.09 = 0.10$. If the moveable scale slides 0.02 unit, then the second Vernier mark will line up with the second fixed mark, and so on. The point is that the human eye is very good at picking out aligned marks. In figure 3.12 you will have no difficulty seeing that the hexagonal nut is 0.76 unit wide. This ability to pick out lines is sometimes called *Vernier acuity*; it was exploited to some extent in earlier instruments of navigation, such as the octant, where a horizon and a star were lined up.

Vernier scales work well only if the graduation marks are spaced very accurately. Accuracy in the machining of scales came 150 years after Vernier developed his scale. In the 1770s Jesse Ramsden, a maker of mathe-

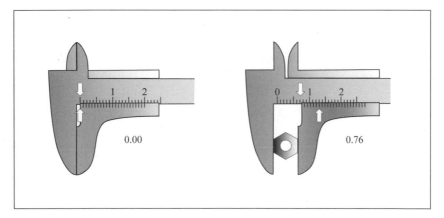

FIGURE 3.12. The Vernier scale. Here, modern calipers show how the Vernier scale is used in practice. The fixed-length scale is at the top and the moveable Vernier scale underneath. It is easy to see which graduation marks line up. In this case, lengths are graduated in units (say centimeters) and tenths of a unit: by adding a Vernier scale, lengths can be measured to hundredths of a unit. Of course, the measurement is accurate only if the caliper machining is good, with tolerances less than a hundredth of a unit.

matical instruments from the north of England, constructed a *dividing engine* that very accurately graduated linear and circular scales. The design of Ramsden's machine was influenced by earlier work, but his dividing engine set new standards of accuracy and was widely copied. It is no coincidence that the great improvement in accuracy of octants and sextants happened during this period.

SEXTANT

Sextants operate on the same principle as octants. They were an improvement because they cover a wider range of angles: ⅙ of a circle instead of ⅛. This became important in the last decades of the eighteenth century because by then navigators had figured out a way to estimate longitude via the lunar-distance method, as we will see. From the relative separation of sun and moon, navigators and geodetic surveyors could consult the Nautical Almanac, determine local time (relative to Greenwich Mean Time, which applied at the prime meridian), and thus estimate their longitude. Sometimes the moon and sun could be separated by more than 90°, however, and so the sextant was developed. The doubling of angles due to mirrors meant that sextants could be used for separations as great as 120°.

Astronomical sextants have been around since the sixteenth century, but sextants that were suitable for use on board ships were invented much later, in 1757 by Captain John Campbell, with the help of a London instrument maker named John Bird. As manufacturing processes improved, the sextant became more sophisticated and more accurate. Silvered glass replaced polished metal mirrors. Filters were added so that the navigator or surveyor could sight the sun. Sextants are still used as backup instruments today (fig. 3.13): improved optics and reduced machining tolerances make modern sextants accurate to less than a tenth of a minute (an error of one minute of arc translates into a position error of one nautical mile).[8]

Sextants are delicate instruments and easily thrown out of alignment— for example, by being dropped. For this reason they are treated carefully and are provided with a protective case. As a consequence, many old instruments survive in good condition. The National Maritime Museum in London has an extensive collection of old navigation instruments, and it is interesting to note, from their dimensions and from the graduation marks, the potential accuracy and the size of each type of instrument. Two trends

8. For the early navigation and surveying instruments (mariner's astrolabe, cross-staff, quadrant, octant, sextant) see Boorstin (1983), Kemp (1976), Mörzer Bruyns (1994), Turner (1988), and Turner (1998, pp. 39–80).

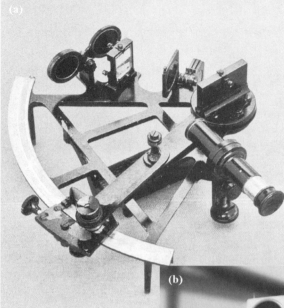

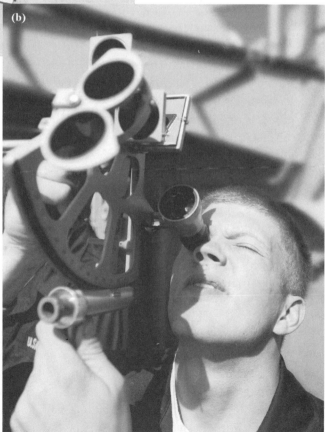

FIGURE 3.13. (a) A sextant. (b) Shooting the sun with a sextant; note the deployed filters. (a) Image courtesy of the National Oceanic and Atmospheric Administration; (b) U.S. Navy photo by Photographer's Mate Third Class M. Jeremie Yoder.

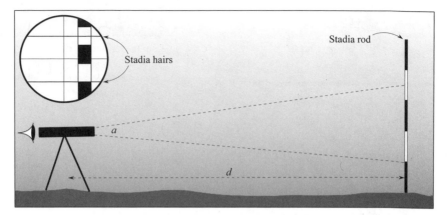

FIGURE 3.14. Measurement of distance using a stadia rod. In this example, the stadia hairs in a sighting scope are separated by angle a = 0.01 radians, and a 5-m stadia rod is placed a distance d away. We observe through the scope that 2.5 m of the rod appear between the stadia hairs, and so the distance d is 250 m.

emerge from a comparative study, as instruments developed over the centuries: increasing accuracy and reduced size. Thus, a fifteenth-century cross-staff might be 90 cm long, while a mariner's quadrant from 1680 has a radius of 75 cm. A backstaff from 1740 has a maximum dimension of 63 cm. An octant from 1785 has a radius of 41 cm.

Old designs, and old instruments, depended upon size for accuracy (more space to graduate the scale). An octant from 1943 has a maximum dimension of 20 cm, though both it and its much larger predecessor from 1785 are calibrated to one minute of arc. A sextant from 1790 with a one-minute Vernier scale has a maximum dimension of 49 cm, while one from 1940, also with a one-minute Vernier, has maximum dimension of 24 cm. Smaller-size instruments are easier to use; thus, there is an incentive to reduce dimensions, so long as accuracy is not compromised. Miniaturization predates the electronic age.

BY DISTANCE MADE MORE SWEET

You will have noticed that most of the development in navigation and surveying instruments we have considered thus far concern increasingly accurate angle measurements. Accurate measurement of distance lagged (it was not so important for maritime navigation) and would not really catch up until the modern era. Before laser rangefinders came on the scene in the 1970s, for many generations distances were measured via *stadia rods* (fig. 3.14). A sighting telescope contained stadia hairs as well as cross-hairs, and these were positioned a fixed angular distance apart, such as 0.01 radians. The stadia rods were marked prominently in such a way that, when a rod was fixed into the ground some distance away from the telescope, the length of the rod that lay between the stadia hairs could be estimated—the idea is shown in figure 3.14—and the distance to the rod could be calculated.

Meet the Theodolites

Land surveying is both easier and more difficult than surveying or navigating at sea. It is easier because the surveyor does not have to deal with a pitching and rolling deck, and because he can use larger and more sophisticated instruments. It is more difficult because he is asked to do more: he must measure in three dimensions—azimuth (horizontal angle), elevation (vertical angle), and distance—and usually without a horizon to help him. Yet such was the inclination (now that's almost a pun) to survey tracts of land with precision, that the development of surveying instruments led, even before the electronic age, to accuracies that were astounding.

In essence, a theodolite consists of a pair of perpendicular sextants on a tripod, from which a plumb line is suspended to the survey point. Consider a tripod that provides a horizontal, circular platform around which a sextant can rotate, measuring azimuth angles. A second sextant is fixed atop this platform and is free to rotate in the vertical direction to measure elevation angles. The theodolite shown in figure 3.15 is a 300-pound monster that dates back to 1836, when it was the pride and joy of the United States Coast Survey.[9] Its base circle has a diameter of 30 inches; this dimension was used in those days to describe a theodolite because size was an indication of accuracy. The theodolite of figure 3.15 produced an error of one part in 100,000.

The theodolite concept is quite old, belonging to the sixteenth century. As technology improved, so did theodolites, which continued to progress until the end of the twentieth century, when they began to give way to

9. This organization was inaugurated by President Thomas Jefferson in the first decade of the nineteenth century to perform an accurate survey of the U.S. east coast, though it took a decade or two to get going. In the United States and other countries at that time, coastal surveys were carried out mainly to decrease the number of shipwrecks. Jefferson was very interested in surveying—he commissioned the Lewis and Clark expedition, one purpose of which was geodetic surveying—and he owned a theodolite that had been made in England by Jesse Ramsden, of dividing engine fame.

satellite positioning technology. Modern theodolites incorporate electro-optics, plus digital data storage and display. Some modern instruments incorporate electronic distance-measuring devices, typically infrared laser rangefinders that can ascertain the distance to a target up to 20 km away, to within a few millimeters. The combination of an electro-optic theodolite and a rangefinder is termed a *total station*. (For a 1950s theodolite and a total station, see fig. 3.16.) Perhaps surprisingly, the best theodolites are slightly more accurate than the best total stations (at the time of publication), and total stations are more accurate than the GPS system. It is a matter of ease and speed of use. A total station can measure angles and distances very quickly and then download the data into a database to be integrated with other measurements. It can be operated remotely and can interface with other systems such as GPS. And with an error of two parts per million (up to a range of 1,500 m, where the angular error is equivalent to 3 mm), we have reached an accuracy that is plenty good enough for most

FIGURE 3.15. "The Great Theodolite." This heavy-weight instrument was state of the art in 1836 when it was acquired by the U.S. Coast Survey, headed up by Ferdinand Hassler. Hassler designed the theodolite himself and had it manufactured in England. The Great Theodolite was used constantly between 1836 and 1873, when it was damaged beyond repair in Georgia during a tornado. Image courtesy of the National Ocean and Atmospheric Administration. This organization can trace its roots back to the Coast Survey.

FIGURE 3.16. A theodolite in action, ca. 1950. This instrument attained an angular resolution of 0.2″, which is equivalent to about 1 cm at 10 km. Put another way, it could distinguish two adjacent dimes from a distance of 6 miles. Inset: A CST-205 total station, with 5″ angle read. Thanks to Coastal Instrument and Supply, Palm Bay, Florida, for permission to reproduce the inset image.

surveying purposes. Total stations are used by archaeologists, crime scene investigators, and miners, as well as by land surveyors.

Global Positioning System (GPS)

Historians now speak of the rapid and all-embracing spread of information technology that has characterized the last quarter century or so as the "third Industrial Revolution."[10] For me, two components of this latest

10. The first Industrial Revolution occurred in England, picking up pace from a slow start in the middle of the eighteenth century and accelerating to its frenetic peak in the first half of the nineteenth. The second Industrial Revolution took place in the rest of Europe and the New World between about 1850 and 1920, and was characterized by a huge increase in manufacturing industries (like the first) plus the electrical and chemical industries. The third revolution arose from the marriage of increasingly sophisticated computer and communications technologies in the 1970s and 1980s.

phase of mankind's technological advance stand out: the Internet and GPS. You may be one of those people who would place cell-phone technology ahead of GPS in terms of the everyday impact it has had on your life, but maybe I can persuade you here that GPS has, or will have, a greater impact further down the line.[11]

The Global Positioning System began life in the 1970s as a military navigation and targeting system and, indeed, is still under the umbrella of the U.S. Department of Defense. The idea draws on the LORAN radar system of World War II, in which radar beams guided allied bombers to their targets in Germany, as we will see. The scope and scale of the GPS concept was much bigger, however: every point on the face of the earth would be covered, and the position of every GPS receiver that is visible from space would be determinable with unprecedented accuracy. This would be achieved via a constellation of 24 satellites, each of which would transmit information about its current position and time.[12] A receiver on or above the earth's surface would need to be able to pick up at least three such satellite transmissions at any instant in order to calculate its position from the information; four would be better. The clocks on board each satellite would have to be accurate to within a nanosecond and be stable for years. The satellite orbits would have to be controllable from earth, and their transmissions would have to be unjammable.

The satellites are dubbed NAVSTAR, which stands for "Navigation Satellite Timing and Ranging"; one is shown in figure 3.17. The first was launched in 1989, and enough of them were in place by the time of the first Gulf War for the growing system to be of significant military value. By the mid-1990s all 24 of the satellites were in place; this is the number needed to ensure that every GPS receiver on earth can obtain information from at least four satellites at any one time. (The actual number varies between 4 and 12.) The satellites are all at the same altitude, 20,200 km above the surface. At this distance, the orbital period of each is 12 hours, so each passes over a given spot on the surface twice a day. Currently, there are 30 NAVSTAR satellites in orbit: 24 operational and 6 spares. The life expectancy of each is about a decade.

The constellation is controlled from Schriever Air Force Base in Colorado Springs and from four monitoring stations across the world. As the

---

11. For a description of GPS see El-Rabbany (2002); for a user's guide see, e.g., Letham (2008); for a technical reference in the context of surveying, see Leick (2004). The official GPS website at www.gps.gov provides a useful overview of all aspects of the system.

12. For example: "It is exactly 9:42 a.m., and I am latitude A, longitude B, and altitude Z."

FIGURE 3.17. Artist's impression of a NAVSTAR GPS satellite. Image courtesy of NASA.

satellites pass overhead twice a day, their status is monitored, their clocks are synchronized, and if necessary, their orbits are tweaked so that they are in exactly the right position. The information that they provide to civilians is free; we pay only for the receiver. Until the year 2000 this information was intentionally degraded, so that a civilian receiver's position would be given to within 100 m or so of the true position. However, such was the usefulness of GPS to civilians that the deliberate degrading of data was disabled,[13] and nowadays 80% of GPS receivers are civilian. Most navigation around the world and an ever-increasing amount of surveying work are performed or enhanced by GPS.

Apart from impressive positioning accuracy, the great benefit of GPS is its simplicity for the end user. Most of the effort and expense was put into the transmitters; the receivers can be hand-held, are relatively inexpensive, and are both quick and easy to use. To take just one example, we saw in chapter 2 some of the benefits of GPS for broad-coverage geodetic surveys: there are practical secondary consequences of this new capability that will increasingly influence our lives, as we will see in chapter 8—improved earthquake prediction, better flood monitoring, improved traffic manage-

13. The Federal Aviation Authority lobbied for undegraded GPS data for airliner navigation, rescue services, fishing boats, land surveyors, hikers, taxis driving science writers from Manhattan to Brooklyn, geodesy research, geologists, geographers, civil engineers, and urban planners—all of whom have benefited significantly from access to GPS data.

ment, improved farm planning and management, and more effective disaster relief.

The accuracy of the system depends mostly upon the quality of the receiver. We have seen that the satellite transmissions are tied to very accurate clocks, but such clocks are very expensive, and the clocks in GPS receivers are much less accurate. The degree to which clocks are accurate influences the accuracy of position fixes, which are attained from trilateration calculations. (I will unpack trilateration for you later.) Receiver clock error can be cancelled out, leaving an error that depends upon the sophistication of the receiver software.[14] The accuracy for entry-level GPS receivers is about 5 m in latitude or longitude. This can be improved upon by fancy (and more costly) signal processing, so that the best GPS fixes can position a receiver to within a few cubic centimeters. (The error in altitude estimation averages about 1.6 times the error in latitude and longitude.)

From the physicist's point of view, GPS is very interesting because it is perhaps the only everyday technology that requires its designers to take cognizance of Einstein's theory of relativity. There are two strands to relativity, and both of them influence GPS: effects that arise from the speed of a GPS satellite relative to a receiver, and effects that arise from the mass of the earth. The first strand is called *special relativity*; it is the simpler consequence of relativity and was discovered by Einstein as a young man (aged 26) in 1905. It is responsible for *that* equation, and it would probably have been discovered within a few years anyway, even without Einstein. The second strand is called *general relativity*. It was the brainchild of Einstein alone, and it is altogether more complicated. He discovered it a decade or so after special relativity, once he had learned the language of differential geometry that is spoken by general relativity.

A GPS satellite travels at a speed of 14,000 km hr^{-1} relative to a receiver

14. Three satellite signals provide a fix, but the receiver clock error can be such that this fix is off by hundreds of meters. This is because the distance of the satellite from the receiver (required for trilateration, as we will see) is determined by timing the signal as it travels between satellite and receiver. A receiver clock error of, say, 1 microsecond, means a position error of 300 m. Because there are four satellites within range at any given instant, a second fix (involving data from the fourth satellite) can be used to cancel the error; the *difference* in arrival times of the signals from the satellites cancels the receiver's clock error. This form of GPS signal processing, known as *differential GPS*, leads to increasingly accurate positioning data. Other errors—notably the varying time taken for a signal to traverse the ionosphere—can also be mitigated by other differential GPS processing techniques.

on earth. According to special relativity, moving clocks tick more slowly. If this theory were not taken into account, a NAVSTAR satellite clock would fall behind ours by about 7 microseconds (μs) per day. On the other hand, general relativity says that clocks close to a massive object (the earth) tick more slowly than clocks that are farther away, and so the NAVSTAR clocks, at an altitude of 20,200 km, should speed up compared with ours by about 45 μs per day. The net result is that if the designers of GPS did not account for both forms of relativity, the satellite clocks would gain 38 μs every day compared with our clocks on earth. That translates into a position error that increases by more than 11 km per day, which would render GPS useless. In fact, NAVSTAR clock rates are slowed down to cancel this effect. Software in the receivers accounts for the special relativistic changes that arise because of the difference in speed of a satellite as it passes overhead. (The 14,000 km hr^{-1} figure is an average; the speed relative to a receiver depends upon satellite position above the receiver.)

Thus, most of the many sources of error for GPS have been anticipated and eliminated, or else can be mitigated, resulting in an accurate and easy-to-use system. The advantages for surveying are clear: GPS does not require a line of sight between two points, and it works at night or in conditions of poor visibility. It can monitor large areas of land at any one time. GPS has its limitations: it degrades when survey points are beneath a dense canopy of trees, and it does not work in tunnels or mineshafts. Consequently, theodolites and total systems will be with us for some time, and yet I hope it is clear from this brief summary that GPS has changed the game. Surveying and navigating are fundamentally different from what they were even 30 years ago; the difference is greater than that between 1980 and, say, 1680.

Surveying Techniques

So much for the tools of surveying. Now let's look at the techniques that surveyors adopt when using these measuring instruments for the different types of surveys that are carried out.

The purpose of surveying is to determine accurately the position of points on the earth's surface, and the angles and distances between these points. The application of such information is varied. Traditionally, it has been used to make maps of coastlines, establish land boundaries, and detail the assets of newly conquered lands for tax purposes. To carry out their tasks, surveyors need to know about geometry and law, and also about engineering and physics.

TYPES OF SURVEYS

We have seen that the ancient Egyptians used peg and rope to measure out land. They did this regularly to reestablish boundaries after the annual Nile flood washed out old markers. The professional Roman building and road surveyors (the *gromatici*) and the land surveyors (*agrimensores*) used the groma to mark out military camps, distribute lands to army veterans (land in conquered territory was the standard reward for service), and survey new lands so that the emperor could establish a tax register. The caliphs similarly employed surveyors, with their alidades and astrolabes, to map the vast new lands acquired after the rapid Islamic expansion of the first millennium. William the Conqueror's surveyors compiled the Domesday Book—essentially a list of assets covering the whole of England. Later, Napoleon instigated cadastral surveys of his newly acquired territory. Again, the purpose was mainly to ascertain the assets of the land of which he now found himself master, for the purposes of taxation.

In addition to surveys for such legal or financial appraisals, for the last four centuries there have been accurate mapping surveys. We have seen something already of the first significant one—that of the Cassinis in eighteenth-century France. Others followed in all the countries of Western Europe and the New World in the eighteenth and nineteenth centuries. The tools were compass and chain, then transit and tape,[15] and latterly, theodolite and electronic distance measurer (EDM), or total station. Mapping surveys may be topographic, or they may be small-scale *plane surveys* (in which the local ellipsoid of the earth can be considered a flat plane) that look at lot boundaries—for example, as a precondition for a house loan. Some surveys are geological; others are for engineering purposes, prior to constructing building foundations, a tunnel, or a highway. Depending upon urgency, scale, and budget, a survey today may be conducted by one person on foot with a total station, or by a team of helicopters with laser scanners. In the latter case, the helicopters' positions are fixed by GPS.

15. A *transit* is a type of theodolite in which the telescope can turn vertically over 180°, so that without moving the base a surveyor can see behind as well as in front. This is useful for traversing new territory, as when laying out new roads or rail lines. The *tape* is the familiar steel tape, usually 100 feet long in the English-speaking world and graduated in units of 0.01 feet.

TECHNIQUES

There are four basic surveying techniques, although the division of surveying into these techniques is breaking down these days as GPS fundamentally changes the field. So, perhaps I should say that prior to 1990, there
were four well-recognized methods for surveying a tract of land: *leveling*,
traversing, radiation, and *triangulation*. The first is simply the use of a level,
a theodolite, and trigonometry to measure the elevation of points on the
surface—typically to provide contours on a topographic map. A traverse
covers a route, say for a highway through hills: it is a preliminary survey
that links points along a line. Distances and directions are measured. An
open traverse will start at A and end at B, whereas a closed traverse will
start and end at A. Radiation surveys establish radiating lines of measured
length and direction from a point within a boundary. On a plane table the
boundary can then be set out on a map. The most common and most
sophisticated of the traditional surveying techniques is triangulation, and I
will spend some paragraphs describing this method for covering large areas
of land, and also its successor, trilateration.

Some of the traditional tools and methods for conducting surveys are
shown in figure 3.18, from an eighteenth-century encyclopedia.[16]

TRIANGULATION

Because angles could be measured more accurately than distances before
the electronic age, triangulation became the normal method for surveying
large or small areas of land. Triangulation is the technique of locating
position by measuring angles and using plane geometry (or spherical geometry, in the case of geodetic surveys). This technique requires a *baseline*
—that is, a line between two points that are separated by a distance that is
known very accurately. From this baseline, a mesh of triangles is constructed (fig. 3.19), and geometry then is used to determine the position of
survey points (the vertices of the triangles).

There will be an error in measuring each angle, and this converts into a
distance error. Fixing the location of a point therefore becomes an exercise
in statistics, as suggested in figure 3.20, given the many data points of a
triangulation network. The errors are not all independent: for example, the
position of point A in figure 3.19 depends upon the errors in the position of

16. Pre-GPS surveying techniques are discussed in, e.g., Davis (1998) and Pugh (1975).

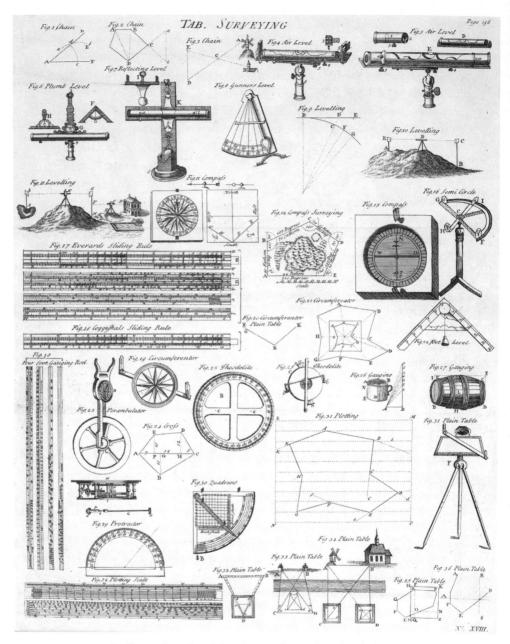

FIGURE 3.18. An illustration of traditional surveying tools and techniques taken from a 1728 English publication.

all the triangulation points between A and the baseline. The statistics of triangulation were worked out by Gauss, who gave us the least-squares method for estimating positions from imperfect data. A triangulation survey yields the least magnitude of error if a *control network*—a mesh of large-scale triangles—is first set up. From this network, points can be filled in inside the control network, a method known as *resectioning*. In this way a coarse framework, a skeleton, of the land area to be surveyed is firmly established, and then fine scale is superimposed—the flesh on the bones.

Figure 3.20 provides a simple illustration of some of the basic ideas of triangulation. Two forest ranger stations, A and B, are separated by a known distance *l*. The line between them is a triangulation baseline. A forest fire is spotted beside a trail at C. The rangers can radio in the direction of the forest fire (angles *a* and *b*), from which the location of the fire can be fixed by triangulation. In figure 3.20 the error in estimating the angles is ϵ; that is, the rangers are assumed to estimate angles to within ϵ of the true angle.

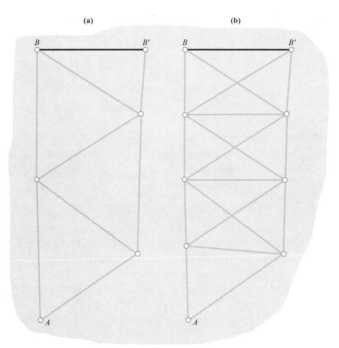

FIGURE 3.19. Triangulation from a known baseline, *BB'*. (a) A chain of triangles leads to the point *A*. From the known length of the baseline and the geometry, the distance from *A* to *B* can be found. (b) Braced quadrilaterals cover the same area. This involves more work but is more accurate because intermediate measurements are cross-checked.

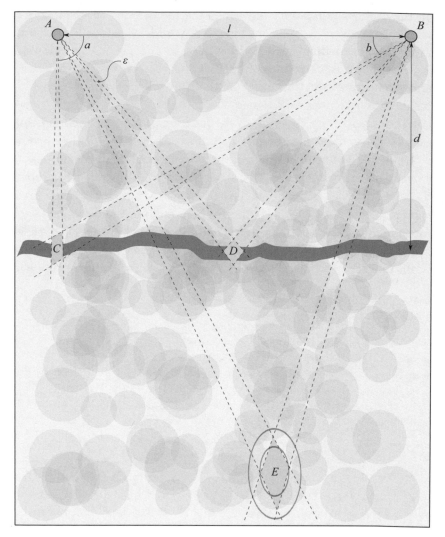

FIGURE 3.20. Triangulation error. A baseline *AB* is formed by two forest ranger stations. The location of a forest fire can be determined by triangulation, but an error ϵ in estimating the angles leads to an uncertainty in the fire location, represented by the shaded quadrilateral boxes *C*, *D*, and *E*. The size and shape of these boxes depends upon the size and shape of the triangles. Let us say that a hiking trail (thick line) runs parallel to the baseline, and so *C* and *D* are both the same distance from the baseline, yet the uncertainty of their positions is different. A detailed statistical analysis replaces the "uncertainty boxes" by ellipses, as shown surrounding *E*, which illustrate that we are, say, 50% confident that the fire is within the inner ellipse and 90% confident that it is within the outer ellipse.

Because of the uncertainty of the angles due to measurement error, there is an uncertainty about the forest fire location, shown as the shaded area surrounding C. Note that the size of the "uncertainty box" varies with position: another fire at D has a different-shaped box, even though D is the same distance, d, from the baseline. (The error is a minimum when the angle ADB is 90°.) A fire farther away at E is surrounded by a much larger "uncertainty box." This example shows how the errors of triangulation depend upon the shape and size of the triangles.

Surveying by triangulation is a very old practice. It may have existed in Han China in the third century BCE. It certainly was used in Arab lands following the Islamic expansion in the seventh century CE, and from the Arabs it came to be known in Christian Spain before the Reconquest. From Spain, however, the knowledge diffused to the rest of Europe only slowly. The idea was proposed as a method for mapping by the Dutch cartographer Gemma Frisius in 1533; eighty years later the modern systematic form of triangulation was worked out and practiced by the Dutchman Willebrord Snell, whom we met earlier. He showed how, in large-scale geodetic surveys, the curvature of the earth could be taken into account. Modern triangulation can turn meshes of triangles into three-dimensional maps displayed on computer screens, showing topography projected onto a screen like a photo. For such cases there is a best choice of triangle vertices (i.e., of survey points). Triangulation with such a set of points, with a view to converting the data points into a 3-D image, is known as *Delaunay triangulation*.[17]

TRILATERATION

Whereas triangulation estimates distances via angles, trilateration does the reverse, estimating the angles or positions of a point given certain distances. It is the method that has to be adopted by GPS receivers, since GPS provides detailed distance information and no explicit angle information. I will outline trilateration here in the context of GPS position estimation. The math of trilateration is complicated, but the basic idea is geometrical and quite easy to convey.

Consider first a two-dimensional trilateration problem. In figure 3.21a we have three satellites at locations A, B, and C. They transmit data to a receiver, and from this information the receiver software is able to calculate

17. Most surveying textbooks include overviews and detailed analyses of triangulation. See, e.g., Davis (1998), Duggal (2004), Kavanagh (2008), Liu (1997), Petrie and Kennie (1987), and Pugh (1975).

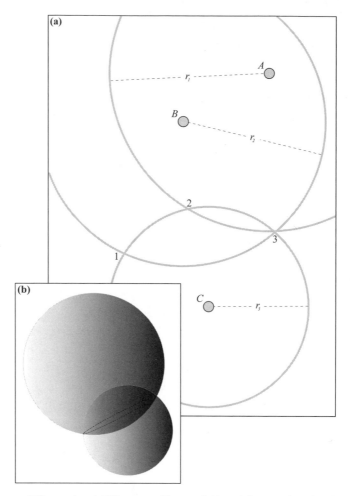

FIGURE 3.21. Trilateration. (a) Three satellites at *A*, *B*, and *C* transmit a signal that tells a receiver it is a distance $r_{1,2,3}$ from them. For one satellite, the distance information is sufficient to show that the receiver lies on a circle centered on the satellite position. For two satellites, two intersecting circles restrict the receiver to the intersection points (here, points 1 and 3, if the two receivers are *B* and *C*). For three satellites, the receiver position is determined unambiguously. (b) In three dimensions, four satellites are needed for unambiguous positioning of a receiver. With one satellite, we can place the receiver on the surface of a sphere; with two satellites we can place it on the circle where two spheres cross (as shown). From (a) you can now see why a total of four satellites are required.

the distance to each satellite, $r_{1,2,3}$. Let us say that satellite B is the first to move into range of the receiver. From this data, the receiver knows something of its own location: it must lie on a circle of radius r_2 centered on B. Next, satellite C moves into range and transmits its data. The receiver now knows it is a distance r_3 away from C and so must lie at one of two positions (labeled 1 and 3 in fig. 3.21a). To fix its position without ambiguity, the receiver needs a third satellite. When A appears and transmits data, the receiver can finally conclude that it is at position 3. (Here for ease of explanation I have assumed that the three satellites appear sequentially to the receiver; in practice their transmissions are monitored simultaneously.)

Thus, for the two-dimensional case, we see that three satellites are needed. The real world is three-dimensional, however,[18] and in this case four satellites are required to uniquely fix receiver position. You can see why this is the case from figure 3.21b. A receiver knows that it lies a certain distance from a satellite, so it lies somewhere on the surface of a sphere centered on the satellite. For two such satellites, the receiver knows that its position must lie on the circle of intersection, shown in the illustration. From a circle we know already that it takes two more satellites to fix the position; thus, a total of four satellites are needed in the real world.

For a receiver on the surface of the earth, we already know one of the spheres, the earth itself (OK, so it is not quite a sphere, but the idea still works), and so you might think that we need only three satellites to fix position on the earth's surface. Technically true, but in practice the fourth satellite is used to eliminate clock error in the receiver, as discussed earlier. In fact, the more satellites that a receiver can detect, the better, so far as positioning accuracy is concerned. The estimated position of the receiver becomes more accurate with each additional satellite it sees because various errors are canceled or partially canceled by the differential GPS technique.

18. Einstein may disagree.

CHAPTER FOUR

Mapmaking

After a brief historical excursion, we look at the problem of map projections: how to represent the three-dimensional surface of our planet on a piece of paper. We will see that there is no way to do so with perfect accuracy and that, as a consequence, many different map projections have been promulgated over the centuries, to enhance this or that feature of the earth.

The First Two Thousand Years

The long and winding story of humankind's struggles to develop accurate maps is in many ways a sorry tale, at least in the West. After a bright start, it gets stuck. Misconceptions were accepted and then became enshrined for millennia. My account of early mapmaking is, metaphorically, shown in large scale—covering a great area but with little detail. There is a danger in this approach: if I pick out a few key events and skip over many others, you may join the dots in a linear fashion. You might then quite naturally consider that mapmaking progressed in an inexorable steady stream, like a rolled carpet unfolding. It didn't happen like that. For every influential figure I mention, there were a hundred others. Many of them doubtless made more telling contributions or were less wayward in their understanding of mapmaking. Space precludes a full discussion (an irony, I suppose). I cover a few key subjects of which we have a reasonable knowledge and omit others of which we don't or which are controversial. As always, I will point those readers who yearn for more detail toward references that provide such.

In the West, cartography starts plainly enough with the Greek philosophers. Later, it becomes entangled with aspects of navigation, exploration, the geographical expansion of nations, and mercantile interests. It becomes heavily influenced by external factors, such as religious orthodoxy and the development of printing. But the beginnings are simple and clear.[1] Let us start with Herodotus in the fifth century BCE. An outstanding historian, Herodotus speculated about the geography of the lands he wrote about. In particular, he knew of the Nile and its annual flooding and wrongly speculated that this must be due to snow melting in mountains very far to the south. Here is an early example of a myth that was per-petuated for millennia; well into the late nineteenth century, European explorers were seeking the source of the Nile too far south. Eratosthenes, whom we met earlier measuring the radius of the earth, proposed a grid system for drawing maps. This idea was improved upon by Hipparchus of Rhodes. In the second century BCE he proposed a reference grid that was based upon astronomical factors, so that, for example, cities on the same line should have the same number of daylight hours during their longest day of the year. Hipparchus is thus credited with giving us the idea of latitude and longitude. He also was the first to write about the precession of the equinoxes.

The contribution of Hipparchus was recognized by a figure who is uni-versally known as the father of geography: Claudius Ptolemy, the official astronomer and geographer of Alexandria. Ptolemy wrote the hugely in-fluential eight-volume *Guide to Geography*, which included a map of the known world (fig. 4.1) with the latitude and longitude coordinates of 8,000 major places. Ptolemy's method was scientific in that he intentionally included enough written information so that others could reconstruct his map, were it to be lost. He established the convention of orienting maps so that north appears at the top. Unfortunately, he also held a number of wrong-headed views that would dog Western opinions for over 1,300 years—such was his authority. Ptolemy believed that the earth was the center of the solar system, that the radius of the earth was some 40% smaller than it really is (based upon a misunderstanding of Posidonius's estimate, as we saw in chapter 2), that Asia spread across 180° east to west (it covers 130°), and that Africa was connected to a so-far-undiscovered great southern continent and so could not be circumnavigated.[2]

1. Early maps and mapmaking are well represented in the literature. Much of this sec-tion comes from Balchin (2004), Berggren and Jones (2000), Boorstin (1983), Campbell (1981), Edson (2007), Harley (1989), Harley and Woodward (1987), and Wilford (2000).

2. Captain Cook was sent to look for this great southern continent. He didn't find it and convinced most people that it did not exist. Consequently, he was the last of a long list of maritime explorers who chased after Ptolemy's will-o'-the-wisp.

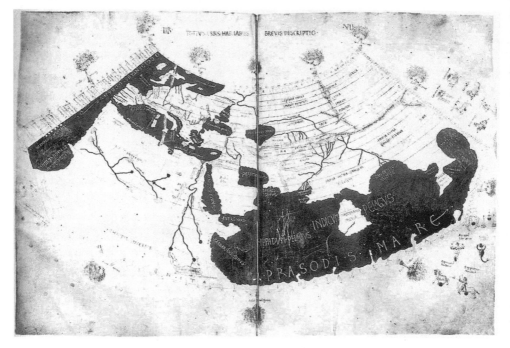

FIGURE 4.1. A fifteenth-century rendition of Ptolemy's world map, from 150 CE. The Mediterranean regions are recognizable; far-flung places like Britain and India are not. The British Library, Harley MS 7182, fols. 58v–59.

The Romans were (perhaps surprisingly) uninterested in mapmaking. Following the demise of their empire, Europe sank into the Dark Ages— dark, at least, in that the corpus of European accumulated knowledge shrank. Little progress in mapmaking was made for a thousand years. (During this period, mapmaking continued unabated in the Orient, and especially in Islamic lands, where knowledge of the Greek philosophers was retained.) Christian dogma claimed that truth was to be found in the Bible, not by scientific investigation, and the maps that were produced in Europe reflected Christians' faith, not the results of exploration or surveying.

Quite a few early maps still exist. The first map that has survived dates from 6200 BCE and appears to be a town plan of Catalhöyük, in modern Turkey. The Egyptians produced maps of larger areas. The oldest known is the Turin papyrus, a topographical and geological map of a 15-km stretch of the Egyptian desert, dating from about 1200 BCE. Anaximander, in the sixth century BCE, was the first Greek who is said to have produced a map, though it has not survived. The Chinese made maps for military purposes and to help with water conservation. In the ninth century Al-Kwarizmi created a world map based on that of Ptolemy, whose method he systemized and whose errors he corrected.[3] This Persian mathematician provided the latitude and longitude of 2,402 localities, and the Islamic regions of his map were, unsurprisingly, more accurately drawn than they were in Ptolemy's map. Later in the same century Al-Biruni (whom we met in chapter 2 measuring the earth) discussed map projection—in particular, projecting a hemisphere onto a plane. He treated measurement errors systematically; in this, too, he was ahead of his time.

The maps of medieval Europe were not intended as representational— that is the kindest way of putting it. More than 1,100 medieval *mappa mundi* (maps of the world) have survived, and they display orthodoxy rather than geographical knowledge. Ptolemy had been translated into Latin from Greek around 1400 CE and was accepted by the Church. The teachings of other Greek philosophers were not, if they made statements that contradicted the Bible. During this period, many of the European world maps were of the *T-O* type, meaning that the land mass of the earth was represented as circular (the *O*), surrounded by water, and separated into three continents by large rivers and seas that formed a *T* (fig. 4.2a). Above the *T* was Asia; to the left of the vertical part of the *T* was Europe, with Africa to the right. The waterways that formed the *T* were the Danube and Nile Rivers and the Mediterranean Sea. Jerusalem was in the middle of the circle (because the Bible referred to it as the center of all things). The T-O maps were oriented with east at the top.

In parallel with the T-O world maps, there built up from the fourteenth century smaller-scale maps of the Mediterranean and Black Sea regions that were intended as practical guides for sailors. The representations of coastlines in these maps were by far the most accurate to this date. These maps were created and continuously revised by the many sailors who plied these waters using the newly developed (for Europe) magnetic compass. The practical nature of these *portolan* maps (the word comes from the Italian for a sailing manual) can be appreciated from the fact that they showed the predominant wind directions for the coastal regions portrayed.

---

3. We get our word *algorithm* from Al-Kwarizmi's name. Around 830 CE he wrote an influential treatise on mathematics from which we derive our word *algebra*.

FIGURE 4.2. Early European maps. (a) A twelfth-century diagrammatic T-O map. Orthodoxy was more important than accuracy. (b) The Catalan Atlas of 1375. For the first time in over a thousand years, Europeans were beginning to produce maps that rivaled those of Ancient Greece. (a) From *Etymologies* by Saint Isadore, Bishop of Seville. (b) From Wikipedia.

The mapmakers, historian Daniel Borstin notes, "found little that was useful in all the speculations of Christian theologian-cosmographers. But they gradually incorporated the piecemeal everyday findings of working mariners" (1983, p. 147). The impressive Catalan World Map of 1450—most definitely not a T-O map—was based upon earlier portolan maps. The earlier Catalan Atlas of 1375, now in the Bibliothèque Nationale de France and shown in fig. 4.2b, illustrates how progress had been made; note the Iberian connection, soon to change the world.

Parallels and Meridians

We have seen how latitude and longitude have been used from the time of Hipparchus to describe a location on the surface of the earth. We have seen that the earth is not exactly a sphere, as was assumed by the ancient Greeks, and we will see how the location of the prime meridian has been a bone of contention. Despite these hiccups, latitude and longitude are so useful that they persist to this day. The earth is very nearly a sphere—we will treat it as such in this chapter—and the prime meridian is now universally recognized to pass through the Royal Observatory in Greenwich, London.[4]

Now it is time for us to investigate how the globe is mapped, so we need to be more precise than earlier about specifying positions on its surface. In figure 4.3 you can see how a point on the surface of a sphere is uniquely determined by two angles: latitude (how far north or south of the equator) and longitude (how far east or west of the prime meridian, which by definition is located at 0° longitude). Lines of constant latitude are called *parallels* because they form circles which lie in planes that are parallel to each other, as you can see from figure 4.3b. Circles of latitude decrease in radius as we move away from the equator. Lines of constant longitude are called *meridians*, from the Latin for "midday," because the sun is directly over a given meridian at midday. Meridians describe great semicircles around the earth, and they all meet at the poles (fig. 4.3c).

To make a map, we need to find a way to translate these three-dimensional coordinates onto a two-dimensional piece of paper.

Globes and Maps

Or do we? Why bother? Why not make 3-D representations of the world? There are a number of advantages of globes, after all. We will see soon enough that 2-D representations of the earth all suffer from distortions of one kind or another. This is, in fact, unavoidable. It is simply impossible to depict a 3-D surface accurately on a flat map. You can see that this is the case with the following simple, but messy, thought experiment. (Feel free

---

4. When specifying position on the surface with great precision, we must bear in mind the eccentricity of the earth, as we saw in chapter 2. The prime meridian is not quite universally located at the Royal Observatory: GPS places it 100 m to the east.

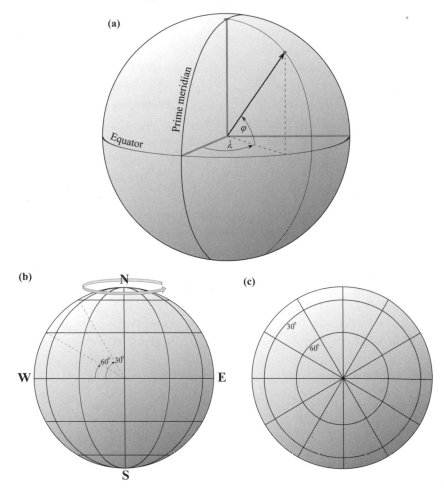

FIGURE 4.3. Latitude and longitude, parallel and meridian. (a) A point on the surface of the earth (here modeled as a sphere) can be uniquely specified by two angular coordinates, latitude φ and longitude λ, measured from the equator and the prime meridian, respectively. (b) Circles of latitude are called *parallels*. Here, the 30th and 60th parallels, north and south, are shown. (c) Viewed from the North Pole, the parallels are circles, and the *meridians* (lines of longitude) are straight lines radiating from the pole.

to turn it into a practical experiment, if you have a globe to play with.) Take a small globe or something approximating it, such as a grapefruit or a baseball. Now take a single piece of paper that is just long enough to wrap around the globe. Cut the paper so that the height is just the same as the height of the globe. You now have a rectangular piece of paper with a length that equals the globe circumference and a height that equals the globe diameter; any mathematician will tell you that your piece of paper therefore has the same surface area as the globe. Yet try to glue the paper to the globe —go on, try it! You may do so without difficulty all the way around the equator, for example, but then you will find that the paper will not reach the poles.

In short, to make a 2-D representation of a 3-D globe, we need to stretch the paper at some points and compress it at other points.[5] This stretching and compressing leads to distortion and is the basic reason why *all* maps are distorted.

So why bother with maps? Let's just make 3-D globes. They have the advantage that they truly represent our earth in miniature. All the directions from one place to another are true; all the sizes and shapes of the continents represented on our globe are true miniature representations of the earth's continents. Unfortunately, the disadvantages of globes outweigh their advantages. Globes are expensive to make, to reproduce, and to keep up to date. They are bulky and so are inconvenient to store and transport. Finally, because they are necessarily small-scale, they cannot provide us with much detail: when was the last time you took a globe on vacation to help you navigate?

Cartographers from centuries ago were well aware of the shortcomings of globes and also of the distortions that resulted when trying to reduce globes to a 2-D surface. Let us examine the various options that exist and look at their advantages and disadvantages so that we can understand why real maps of the world evolved in the way that they did.

Projections

Consider two points A and B on the surface of the earth. These points will lie a certain measurable distance from one another, and to get directly from A to B we travel in a certain well-defined and measurable direction. Consider this distance and direction to have been already determined. Translating the coordinates of latitude and longitude from a 3-D globe onto a 2-D map (*projecting* the globe onto a plane, in the language of geographers) introduces distortions that may result in A and B being separated by the wrong distance, or may result in the direction between them being

---

5. If the paper has the same area as the globe, and if it needs to be stretched to reach the poles, then it has to be compressed somewhere else.

incorrect, or both. Different types of projection can preserve one or other of these navigational figures—distance or angle—but not both.

There is another property that we need to consider: the area of a country or continent. A true representation of the globe will preserve the areas of different regions of the globe. Thus, there are three types of distortion that can arise when a globe is projected onto a 2-D map:

- *Area*. A true map will be *equal-area*, meaning that the relationship between areas on the map is the same as the relationship on the globe. In other words, the map *scale* is true everywhere. For example, South America has an area of 17,840,000 km² (a little under 6.9 million square miles), whereas Greenland's area is 2,166,086 km² (say 835,870 square miles), so South America is about 8¼ times bigger. A 2-D map that is equal-area will preserve this ratio as well as the ratio of any other two areas on the globe.[6]
- *Distance*. No 2-D map can preserve all the distances between two points on the surface of a globe. This will become clear when we look in more detail at map projections. An *equidistant* map is one in which distances in certain chosen directions are correctly represented. For example, an equidistant map centered on New York will show the true distance from New York to London, and from New York to São Paulo—but not from London to São Paulo.
- *Angle*. A *conformal* map preserves the angles between points on the globe and therefore preserves the shapes of areas. In fact, this usually works only locally, meaning over a small area. Thus, if we draw a line from Baltimore to Washington, DC, to Philadelphia, the angle at Washington on a conformal map equals the angle measured on the globe.[7]

6. Does your map of the world show Greenland to be larger than the Democratic Republic of the Congo (Zaire)? If so, your map is distorting areas; the DRC is 8% larger.

7. The angles of a triangle add up to 180°, as you may recall from high school. However, this is true only on a flat plane, such as a piece of graph paper. On a globe, the three angles of a triangle add up to more than 180°. For example, consider a triangle drawn from latitude 0° and longitude 0° (on the equator just off the coast of central Africa) to latitude 90° north (the North Pole) to latitude 0° longitude 90° west (just north of the Galapagos Islands). Each of the angles of this triangle is 90°, so the sum is 270°. (The spheres of fig. 4.5 are divided into eight such triangles.) Smaller triangles on the globe (with each leg of the triangle much smaller than the earth's radius) have angles that add up to a smaller number, much closer to 180° though still bigger. This is why conformal maps preserve angles only locally—over a small area surrounding a given point on the surface—because over such an

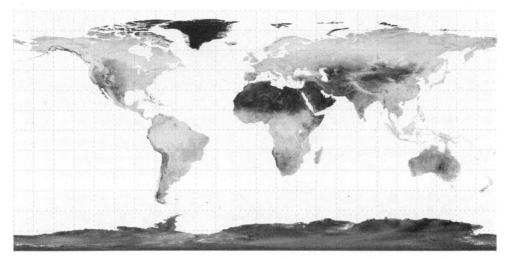

FIGURE 4.4. Unprojected, or *plate carrée*, map of the world. Distances, angles, and areas are all distorted. Such a map is easy to draw, however, because the x-axis (horizontal axis) of the map is just longitude, while the y-axis (vertical axis) is latitude. Thus, parallels are horizontal and equally spaced, and meridians are vertical and equally spaced. NASA image.

It is impossible for a map (meaning, from now on, a 2-D representation of the globe) to represent areas truly *and* represent distances truly *and* represent angles (i.e., shapes) truly. A good map will represent one or more of these properties truly, but not all of them together. For example, no map can be both conformal and equal-area. No map can be both equidistant and equal-area. Because of these different types of distortion, car ographers tell us that there is no single best map projection; the best projection for one purpose may be poor for another. We will see examples soon enough.

First, however, let us consider a map that is *not* projected. Why don't we simply take the latitude and longitude coordinates of the globe, and map them onto a plane? Say the x-axis is longitude and the y-axis is latitude. Such an unprojected map of the world is shown in figure 4.4.[8] It distorts scale and area everywhere except at the equator and distorts more at greater distances from the equator. These distortions, however, are not as

_____

area the surface is nearly flat. The sum of angles of triangles drawn on a sphere thus varies with triangle size. Such is the weirdness of spherical geometry.

8. An unprojected map is also known as a *plate carrée* map, an *equirectangular* map, or an *equidistant cylindrical projection*. The distances between parallels and between meridians is the same everywhere on the map.

bad as those that result from the much more familiar Mercator projection. The map of figure 4.4 also distorts distances except along the north-south direction and along the equator. It also distorts directions except along the cardinal compass points (north, south, east, and west). Because of these distortions, it is rarely used. Its main advantage accrues to the mapmaker, not the map user: it is easy to draw.

Three Common Types of Projections—and an Oddball

There are dozens, hundreds, of named map projections. Why so many? Because each has an advantage over its fellows, in one way or another, as we will see. Maps are used for very different purposes by different people. A navigator in an Age-of-Sail ship, plowing the open oceans, required nautical charts that served a very different purpose from the maps used today to monitor the spread of pandemic diseases. A different map would be better at showing the world's ocean currents; yet another map will be better suited to displaying airline routes; and so on.[9]

CYLINDERS, CONES, AND PLANES

The majority of maps fall into one of three basic families: *azimuthal projections*, *conical projections*, and *cylindrical projections*. These three categories are illustrated in figure 4.5. The trick is to wrap a piece of paper around a globe. We cannot stretch the paper, so we cannot wrap the globe completely. By wrapping the paper around the equator, as in figure 4.5c, we have formed it into a cylinder. From this configuration, if we can map the globe onto the paper by performing a *projection*, then every point on the globe is systematically placed upon the paper, and when we straighten out the paper, we have a map. So how do we perform a projection from spherical globe to cylindrical paper? There are many different ways, as we will soon see.

Conical projections adopt the configuration of globe and paper as shown in figure 4.5b. Paper can be twisted into a cone without stretching it, so this configuration works. We project somehow from globe to cone, unfold, and we have a conical-projection map. For the azimuthal projections, we don't need to twist the paper at all: it remains a flat plane,

9. Snyder (1993) describes almost 200 projections in his technical history of the subject. Readers who crave the detailed math can find it in, e.g., Dent (1998) and Yang, Snyder, and Tobler (2000). A much lighter, humorous introduction is provided by Gonzalez and Sherer (2004).

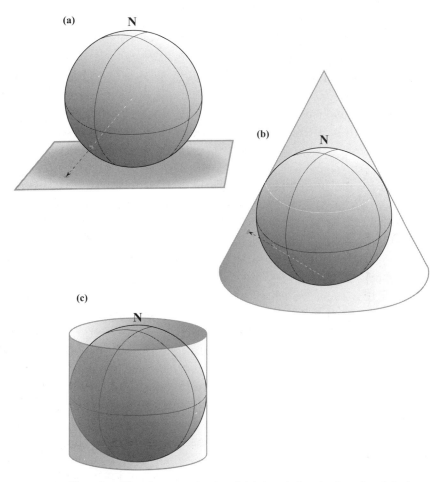

FIGURE 4.5. Three families of map projection. (a) Azimuthal projection: the globe is placed on a flat piece of paper. (b) Conical projection: a paper cone is placed on the globe. In this case, the cone and the globe touch at a parallel that is north of the equator (shown here as a curved white line). The base of the cone is open; a ruler placed across its diameter would just touch the South Pole. (c) Cylindrical projection: a flat piece of paper is wrapped around a globe. In this case, the paper and the globe touch at the equator. Projections are illustrated by arrows for the conical and azimuthal cases. In (b) I have chosen to project from the South Pole, whereas in (a) the projection is from the center of the globe.

as shown in figure 4.5a. In this case, the globe is simply placed on the flat sheet of paper and a projection is performed for the configuration shown. For this reason, azimuthal projections are conceptually the simplest. Therefore, we begin our detailed look at projections by considering these projections onto a plane.

AZIMUTHAL PROJECTIONS

The idea underlying mathematical projections from a point on a sphere to a point on a plane is illustrated by the arrow in figure 4.5a. The geometry is much simpler to grasp than the algebra.

Let us imagine that we shoot an arrow out from the center of the globe shown in figure 4.5a. This arrow will intercept the surface of the sphere at a particular point on its southern hemisphere and then intercept the plane. We say that the point on the sphere has been projected onto the plane. It is obvious from the illustration that every single point on the southern half of the sphere can be projected in this way onto some point of the plane. In an azimuthal projection (which is the kind of projection that Al-Biruni concerned himself with), the projection is a systematic way of sending every point on the globe (that we want to map) onto a point on the flat plane, or sheet of paper.

There is nothing unique about the map center point chosen for the azimuthal projection shown in figure 4.5a. I happened to choose the South Pole, but I could have chosen any other position on the globe as the center point. Thus, for example, if we wanted to map the Arctic regions, we might place the flat plane on top of the sphere, so that the center point—the point of contact between sphere and plane—is the North Pole. For such a map, distortions will be small near the North Pole and will increase further south. In the case of projections from either the South or the North Pole center points, the equator is infinitely far away, so this type of projection is useful only for latitudes near the pole.

Projections from a center on the globe onto a plane are called *gnomonic*, and gnomonic maps have certain advantages. You can see that, for the gnomonic projection of figure 4.5a, Antarctica will be well represented. Near the point of contact, the projection is faithful; distortions increase as distance away from the point of contact increases. Gnomonic projections are good for depicting areas like Antarctica that spread significantly both north-south and east-west.

In addition, gnomonic projections turn great circles into straight lines. It is well known that the shortest distance between two points on a sphere is the line formed by the great circle that connects them.[10] Thus, on maps formed from gnomonic projections, the shortest distance between the

10. Imagine two points, A and B, marked on the surface of a sphere. These points, plus the center of the sphere, define a plane. Imagine this plane cutting the sphere. The

point of contact and any other point on the map is a straight line. This property makes gnomonic maps useful for plotting airplane routes from an airport, with the airport latitude and longitude as the gnomonic center point.

The gnomonic projection has certain disadvantages, as I have indicated. One of those is that only half the globe can be projected onto the plane (i.e., the paper). We can get around this problem by projecting, not from the center of the earth, but from the opposite position on the globe. In the case shown in figure 4.5a, the projection would be from the North Pole, and every point on the globe except the North Pole would be projected onto the plane. Projections like this one, with the projection point placed on the globe in a position that is diametrically opposite the center point, are called *stereographic* projections. In practice they are useful only for regions near the center point, where the map thus formed is approximately conformal—that is, it does not distort shapes or directions very much.

For azimuthal projections (onto a plane) the projection point does not have to be the center of the globe, or a point on the surface of the globe. If we choose the projection point to be infinitely far away from the globe, then the resulting map is an *orthographic* projection, shown in figure 4.6. It looks like a drawing of the earth, because perspective is accurately depicted. Distortions are minimal near the center of an orthographic projection and increase toward the edge. Such distortions are easy for us to accommodate mentally because we recognize that the map is an image of the earth and we can allow for the perspective distortion.

CONICAL PROJECTIONS

Now we turn to projections from a spherical globe onto the curved surface of a cone. This time we figuratively shoot an arrow out from the South Pole (from the bottom of the sphere shown in fig. 4.5b). You can see that the cone is completely filled by this projection: every point on the surface of the cone corresponds to a point on the sphere. All but one point on the sphere is mapped onto the cone. The South Pole of the sphere is unique in this projection because it projects to more than one point: it can be thought of as projecting to every point on the circle that defines the bottom edge of the cone. Unfolding the cone, we then have a flat map.

----

intersection of plane and sphere is a circle—in fact, a great circle, because its center is the center of the sphere. The section of this great circle connecting A and B is the shortest distance between the two points that lies on the surface of the sphere.

FIGURE 4.6. Orthographic projection. This projection looks like a view of the world (or more accurately half the world, mostly the eastern hemisphere in this case) from space. Our minds are easily able to allow for the distortion produced by this type of azimuthal projection. Adapted from a Wikipedia image.

If you think about it, you can also see that the sphere has to be stretched in varying degrees to project it onto the cone. For example, two points on the sphere that are close together, one just east of the South Pole and one just west of it, will be projected onto opposite sides of the cone. The distance between the points will thus be greatly exaggerated. On the other hand, consider the points of contact between sphere and cone: these points form a circle, a parallel at a latitude that depends upon the cone angle. Any point on this contact circle (shown in fig. 4.5b) is not moved by the projection. Consequently, along this circle distances are preserved, and near this circle the shape of the map on the cone will look very much like the shape of the map on the globe. Distortions will increase with increasing distance away from this "circle of contact" between the sphere and the cone.

Here we have a difference between azimuthal and conical projections. For azimuthal projections, the plane is tangent to the globe at only one point, whereas for the cone it is tangent at a set of points that form a circle.

Google Earth

The most popular geographic information systems (GIS) application, without a doubt, is the Google Earth personal computer application. It is a product of the third millennium—a virtual globe that is possible only because of the reams of digital data made available to us from radar satellites and optical eye-in-the-sky satellites and from databases containing all sorts of geospatial data. Most readers will have some familiarity with this computer program. You may already have zoomed in from outer space and hurtled headlong toward the earth, viewed in bird's-eye view until you got very close; then the angle changed, and you appeared to fly across the surface of the earth.

You will by now appreciate the considerable amount of data and data processing that is required to produce a program like Google Earth. The map projection that is used when we initially zoom in toward Earth, looking down from directly above, is called the *general perspective* projection. It is like an orthographic projection, except that the viewpoint is not infinitely far away but is instead at a finite distance, one that decreases as we zoom in. In other words, Google Earth tries to present us with a photographic image of what we would actually see.

The data are arranged in layers, which declare themselves as we zoom in. We can choose additional layers—topographical or demographic details such as roads, political boundaries, store locations. We can change the angle and zoom in or out diagonally and approach a given location—say, Prague in the Czech Republic—from different directions, giving us the impression of a 3-D view. Other data, such as the NASA radar data of the Grand Canyon, are processed using software that generates relief maps and so creates more convincing 3-D views: with Google Earth you can fly along the Grand Canyon. I mention Prague because it happens to be the city that can be viewed with the greatest clarity: it is imaged with a resolution of 0.1 m (most other cities are presented at a resolution of 0.3–1.0 m). Rural land is generally provided with a lower resolution (15 m).

The fusion of several different types of data makes the presentation a little uneven (for example, a low-resolution strip of land abuts a higher-resolution strip and looks quite different), but the potential for the future is apparent: better data will permit improved resolution, with sharper images—and we won't be able to see the join between two datasets. Indeed, this type of improvement is ongoing.

If your computer has a joystick, then Google Earth (which is free) can be used as a flight simulator. In addition to Earth, we can view the moon and Mars.

Local Projections

In the figure we see that projecting from one plane to another does not lead to distortions. Any shape that is inscribed in the top plane is faithfully mapped onto the bottom plane, via a projection from any chosen point above the top plane. All angles and shapes are preserved; the only map feature that changes as a result of this projection is the scale.

Contrast this plane-to-plane projection with a sphere-to-plane projection. In part (b) of the illustration, we have a cross section of such a projection: the sphere appears as a circular arc and the plane as a straight line. Now you can see that distances are distorted. (Distance *AB* equals distance *BC*, but *DE* does not equal *EF*.) However, if the projection is restricted to small regions of the sphere nearest the plane (regions with a linear extent that is small compared to the sphere radius) then the sphere surface appears nearly flat, and so distortions are small. The projection of part (c) resembles that of part (a).

The message to take away from all this is that low-distortion map areas are those in the region where sphere and plane meet—or near the closest point, if they do not meet, as in part (c) of the illustration. The same applies to a cone: in this case, as we have seen, the tangential contact region forms a circle, not a point, and so for conical projections, regions near the contact circle suffer little distortion.

A subclass of projections called *secant projections* increases the area of globe that suffers from little distortion by changing the projection geometry. If, in figure 4.5a, the sphere does not sit on the plane, but instead sinks into it a little, then the contact point becomes a more extended contact circle. Careful choice of geometry (how much the sphere sinks into the plane) can result in small distortions over quite a wide area of the map. Similarly for conic and cylindrical projections: the spheres of figure 4.5b and 4.5c can be allowed to penetrate the cone and cylinder, thus increasing the contact points and the area of the globe that can be projected onto a map with little distortion.

We have seen that there are three different types of distortion: distance, area, and shape or direction. All three of these are small in regions of the globe that are near the point(s) of contact for the projection. For some projections, as we have seen, one or other of the distortions may be reduced or be absent farther away from the points of contact.

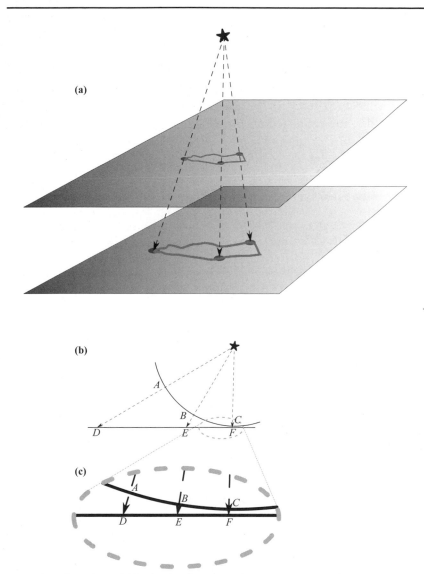

Projection distortions. (a) Projecting from one plane to another produces no distortions. Here, a shape on the top plane is projected from a point (the star) onto the lower plane. All that changes is the scale. (b) Now we project from a point, through a globe (circular arc), and onto a plane (straight line). Distance *AB* equals distance *BC* on the surface of the globe, but the projected image does not preserve distance: *DE* is not equal to *EF*. (c) If, however, we restrict our projection to globe areas nearest the plane, distortion is reduced.

Near contact points, distortion is minimal. (Is it clear to you that near such tangent points, the distortions that are generated by projection are minimized? If not, read the "Local Projections" sidebar.) Thus, conical projections produce a strip of map with minimal distortion. Consider again figure 4.5b. Let us say that the cone angle is chosen so that the contact circle with the globe corresponds to, say, the 50th parallel north. It so happens that much of the populated region of Canada occupies land that runs just north of the 49th parallel—the long border with the United States. So, for places like the border region of southern Canada that stretch long distances east-west and yet are limited in north-south extent, conical projections are good news.

CYLINDRICAL PROJECTIONS

Projections from a sphere onto a cylinder (fig. 4.5c) are as many and as varied as projections onto a cone or a plane. We might project from the center of the globe onto a cylinder, for example, or we might project as in figure 4.7, which shows the *cylindrical equal area projection*, created by Lambert in 1742. The global map that results from this projection is shown in figure 4.8. It looks distorted—and is distorted, but no more so than some other, more familiar projections. It distorts almost everything but has the advantage that it is equal-area everywhere. However, because it is so strange, it is little used today.

There are dozens of other cylindrical projections, the most famous of which is the *Mercator projection*. This projection was very popular with navigators in the Age of Sail because it has one very desirable property: straight lines drawn on the map represent courses of constant compass direction. Such a course is known as a *rhumb line* or a *loxodrome* and was the natural course for a ship of centuries past to follow because in those days navigators had few tools other than a magnetic compass to guide them. A straight line toward the destination on a Mercator-projection map would indicate the true course to be followed. If, for example, a straight line drawn from a ship's current position to the desired destination indicated that the ship should set a northwesterly course, all the navigator had to do was point his vessel to the northwest, as indicated by his compass. Following this course would send him to his desired destination. Clearly, it was very important in those days to have a method of navigation that was both simple and reliable.

The Mercator projection is the only map projection with this straight-line loxodrome property. Meridians and parallels are straight on a Merca-

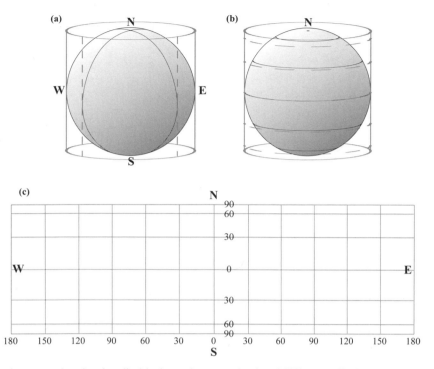

FIGURE 4.7. Lambert's cylindrical equal area projection. (a) Wrap a cylinder around the globe, and cut the globe with a plane, passing through the north-south axis. The plane will intersect the globe along meridians (great circles), and the cylinder along straight lines, as shown. This is how longitude is projected. (b) Cut the globe horizontally with a plane. The plane will intersect the globe along parallels (lines of constant latitude), and the cylinder on circles, as shown. This is how latitude is projected. (c) The unfolded cylinder yields a flat map. Toward the poles, the parallels get closer together. (The opposite occurs for the Mercator projection.)

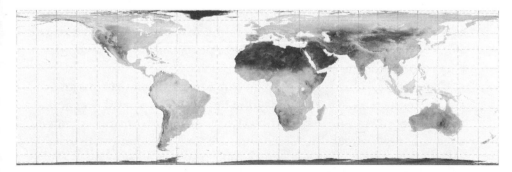

FIGURE 4.8. World map derived from a Lambert cylindrical equal area projection. This map looks very squashed, especially near the poles. Its one saving grace is that it preserves areas, though shapes can be horribly distorted. NASA image.

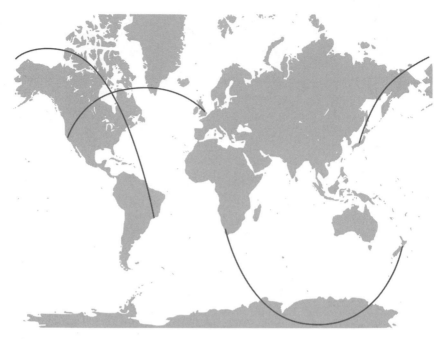

FIGURE 4.9. Mercator's projection, from 80° north to 80° south. Area is distorted, particularly near the poles, but loxodromes are straight lines on this map. Great-circle routes are complicated, however. This illustration shows three great-circle routes marking the shortest distance routes between Wellington in New Zealand and Cape Town in South Africa, between London and Los Angeles, and between Tokyo and São Paulo.

tor projection (except for the so-called *oblique* variant), and meridians are always perpendicular to parallels. Another benefit of Mercator maps is that they are conformal, so that locally, the maps accurately reproduce shapes, distances, and angles. Thus, for example, while Greenland is represented as far too big, its shape is not much distorted. The Mercator projection also works well for equatorial maps (because the projection cylinder is tangential to the globe at the equator).

There is no such thing as a free lunch, however. Mercator distorts areas and distances, more so farther away from the equator and horribly so near the poles, as you can see in figure 4.9. (Recall that South America is more than eight times bigger than Greenland.) Not only that, but great circles are also complicated—far from the convenient straight lines of, for example, gnomonic projections. In figure 4.9 we see three great circle routes (that is, shortest routes) between distant cities. As you can see from this figure, the Mercator projection will not easily get you from *A* to *B* in the shortest time. Yet, it is not difficult to appreciate why the projection dominated cartography and navigation for centuries: straight loxodromes won out over almost everything else in the eighteenth and nineteenth centuries.

The mathematics of Mercator projection is more complicated than the simple geometric point projections discussed above, or the so-called "normal" Lambert projection. In this case, I have made an exception to my general rule of not presenting you with the mathematics of navigation and have provided a derivation of the Mercator projection in the technical appendix. You will see how the straight loxodrome property is built in, and the derivation will give you an idea of the type of math that is involved in mapping, whatever the projection may be. The message to take away from the appendix is that a map projection can be mathematically fine-tuned to cater to a particular need—to minimize a particular distortion (zero angle distortion in the case of Mercator). The price to be paid is an increase in the other types of distortion (area and distance, in the case of Mercator).

The *transverse Mercator* projection rotates the axis of the cylinder 90° so that it is perpendicular to the earth's north-south axis. In this way the polar regions can be accurately represented. In addition, a well-chosen transverse or oblique Mercator projection allows accurate mapping of countries such as Chile, which stretches along a meridian with little east-west extent.[11] Another advantage of transverse Mercator maps is that they can be joined vertically without worrying about changing scale across the maps.

DYMAXION MAPS

Now we turn to the oddball family of projections promised earlier. We have seen that projections from a sphere onto a plane (which well approximate projections from the globe onto a paper map) necessarily introduce distortions. We have also seen that these distortions are reduced in regions where the map (be it a plane, or a sheet rolled into a cylinder or cone) is close to the globe and is tangential or almost tangential to it. Well, there exist in mathematics a group of beautiful 3-D constructions, the regular polyhedra, which can serve to replace the cylinders and cones. True, we cannot simply construct a polyhedron from a flat piece of paper by curling it up, as we did with the cylinder and cone: to make a polyhedron, we have to cut the paper. But it can be done in such as way that the paper is not stretched and remains in one piece, as we will see.

11. A transverse Mercator projection (also known as the *Gauss-Krüger projection*) sets the cylinder angle perpendicular to the earth's axis, whereas an oblique projection, mentioned earlier, sets it at some other angle between 0° and 90°.

Mercator

Gerardus Mercator was born in 1512, shortly after his family had moved to Flanders from Germany. He received an education in Christian doctrine, dialectics, and Latin. The son of a shoemaker, Mercator was able to study at Louvain University only because of the generosity of his uncle Gijsbrecht and graduated in 1532 with a master's degree in philosophy. After a three-year absence during which he seemed to be assailed by religious doubts, Mercator returned to Louvain to study mathematics under the influential Dutch mathematician, astronomer, and cartographer Gemma Frisius, whom we met earlier. Both frequented the workshop of skilled goldsmith and engraver Gaspar à Myrica; these three men made Louvain a center for globe and mapmaking. Mercator seems to have been very good with his hands and learned to make scientific instruments (compasses, sundials), but he earned a living from mapmaking—he was a superb engraver and calligrapher. At age 24, Mercator married Barbara Schellekens; they produced six children, though five of these died young.

An engraving of Gerardus Mercator by Nicholas de Larmessin. *Bibliotheca Belgica* (1739), by Joannis Francisci Foppens.

Mercator's first map (with Frisius and Myrica), of Palestine, was published in 1537 and was an immediate success. He became famous three years later with his accurate Flanders map, for which he surveyed land using the triangulation method (surely an influence of Frisius). He also made terrestrial and celestial globes, in 1541 and 1551.

Mercator was sympathetic to Protestantism and in 1544 was imprisoned for seven months on suspicion of heresy. It was a sign of the intolerant times in which he lived that suspicion about Mercator arose partly because of his frequent absences from Louvain on mapmaking expeditions. The university stood by him, however; when released, he was able to return to his former duties and researches.

At age 42 Mercator moved permanently to Duisburg in what is now Germany and set up a cartographic workshop. Here, he created his most important works. He made maps of Lorraine, now lost, and of the British Isles, before producing a masterpiece, his map of Europe. Fifteen years later, in 1569, he made a world map "ad usum navigation" (for the use of navigation)— the only occasion he employed the Mercator projection.

Mercator, who coined the word *atlas* to describe a group of maps, died aged 82 in Duisburg.

The Mercator projection probably precedes Mercator, and the theoretical underpinning was done by others, notably Edward Wright at the end of the sixteenth century. However, credit for the projection accrues to Mercator because of the excellent quality of the map he made with it. While this map benefited travelers and geographers, through precise measurement and high-quality production (Mercator discarded many of the historical errors that cluttered medieval world maps), the Mercator projection was ahead of its time and would not really benefit navigators until the eighteenth century because marine navigators lacked the means to measure their location at sea. Once the longitude problem was solved and magnetic declination understood, Mercator projection maps permitted ships to sail the world's oceans secure (if not quite safe) in the knowledge that they knew where they were going.*

* Crane (2002) provides a detailed biography and cartographic history. Taylor (2004) portrays the times in which Mercator lived. Shorter biographical sketches can be found in, e.g., *Encyclopaedia Britannica* and *Encarta Encyclopedia*, s.v. "Gerardus Mercator."

The regular polyhedra are highly symmetrical 3-D objects that are made up of a number of regular polygons (equilateral triangles, squares, or pentagons). For example, six squares can be used to construct that most familiar of regular polyhedra, the cube. Four equilateral triangles can be used to construct a tetrahedron; eight make up an octahedron; twenty make an icosahedron. Twelve pentagons form a dodecahedron. The octahedron and the dodecahedron are illustrated in figure 4.10, which also illustrates how they unfold so that they lie flat.

Here lies the utility of such polyhedra. Because a single piece of paper can be cut and folded to form a polyhedron, we can project our globe onto the polyhedron and so form a map. In figure 4.10 you see that we can place a sphere inside the polyhedron so that each face of the polyhedron is tangent to the globe. Clearly, the more faces a polyhedron has (i.e., the more polygons that are used to construct it), the closer it can be made to approximate a sphere. An exotic example of this idea has found practical application in the construction of soccer balls: figure 4.11 shows a *truncated icosahedron*, which is made up of 12 pentagons and 20 hexagons, next to a soccer ball.

From what we have learned about projections and polyhedra, it seems reasonable to conclude that we can construct a map by projecting from the center of the earth onto one or another type of polyhedron. We can expect fairly small distortions from such a gnomonic projection, particularly if the polyhedron consists of many component polygons and approximates the shape of a sphere. Having made the projection, we then unfold the polyhedron to form a flat map. Such a *Dymaxion map*, as these

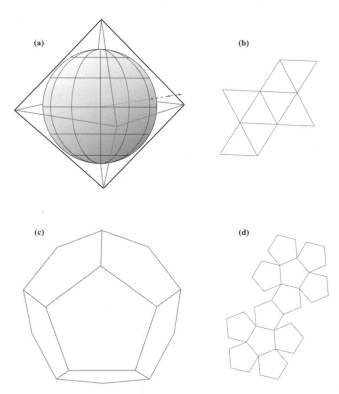

FIGURE 4.10. Polyhedra and Dymaxion projections. In these projections, a polyhedron, rather than a cylinder or a cone, is wrapped around the globe. (a) An octahedron, with eight triangular faces tangential to the sphere. A gnomonic projection is illustrated. (b) The octahedron can be unfolded and laid flat in this way. (c) A dodecahedron, consisting of twelve pentagons, more closely resembles a sphere. (d) This is how the dodecahedron unfolds.

FIGURE 4.11. A truncated icosahedron, constructed from 32 polygons, and a soccer ball. Adapted from a Wikipedia image.

polyhedral-projection maps are called, is shown in figure 4.12 for the case of a dodecahedron.[12]

The figure also shows some of the problems of this type of map. The map shape is odd, and land surfaces can be cut up such that many adjacent geographical features are placed on polygons that are not adjacent. Also, look at the parallels and meridians: they are curved, the lines are discontinuous at the polygon boundaries, and parallels are not always perpendicular to meridians. Another problem is that direction is unclear: which way is north? The Dymaxion maps (also known as *Fuller* projections) are regarded as something of a novelty, not to be taken too seriously, despite their low distortion.

The Menagerie: Other Types of Projections

There are many different types of projections; I have discussed just a few above. Some projections lead to maps with one or another very desirable property that applies across the entire map, though there are few of these.

12. The idea of Dymaxion maps was invented by the American architect Buckminster Fuller and patented by him in 1946. It is natural to use a gnomonic-type of projection for this map, in which the point of projection is the center of the sphere, but Fuller in fact preferred a slightly different form of projection, in which the polyhedron is shrink-wrapped onto the sphere. Fuller reserved the name Dymaxion for the case of an icosahedron, but here I use it to describe any polyhedral projection. For a nontechnical account of Fuller's Dymaxion map, see Edmondson (2007).

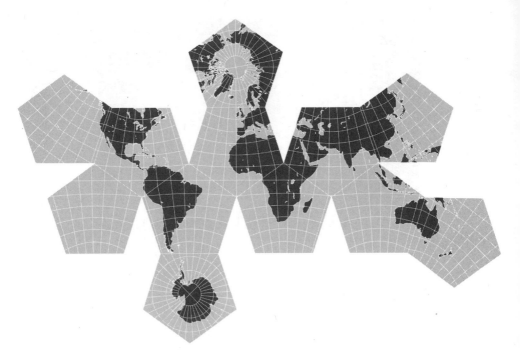

FIGURE 4.12. A Dymaxion-type map, resulting from projection onto a dodecahedron. Which way is up? Thanks to Carlos Furuti (www.progonos.com/furuti/MapProj/Normal/ ProjPoly/projPoly3.html) for kindly creating this image.

For example, there is only one type of conic projection that is equal-area, and only one that is conformal. Other projections produce good properties locally, or along a chosen meridian, or around the equator. Many projections are chosen as a compromise: they are distortion-free nowhere but suffer from low distortion over a wide area. A popular example today is the *Winkel Tripel* projection, which dates back to 1921 (fig. 4.13). It is currently the preferred choice for the National Geographic Society world maps.

One example of a projection developed for a very specific purpose is the *space oblique Mercator* projection, developed by the U.S. Geological Survey. This projection produces no distortion along the ground track beneath an orbiting satellite and only small distortion in a limited swath of the earth's surface close to the ground track. Therefore, a remote-sensing satellite can image the earth beneath it, and the data can be turned into a low-distortion map by using this new type of projection.

The *Miller* projection is a version of the Mercator that is slightly modified to reduce the more obvious scale inflation at the poles. The *Robinson* projection produces a better balance of size and shape at high latitudes and has been used by the National Geographic Society for some of its world maps. *Polyconic* projections, dating to 1820, use different cones at different latitudes to reduce distortion; these projections were once used extensively to map the contiguous United States.

Interrupted projections change the reference point or line in mid-ocean in order to minimize distortions on the continents. These maps look like an orange peel; they are shaped a little like Dymaxion maps but with north at the top of the map. A common example is the *sinusoidal equal-area* projection. Also known as *Sanson-Flamsteed*, this projection dates back to 1570, contemporary with Mercator. We have seen that equal-area maps are good for illustrating distribution patterns, and the sinusoidal version is widely applied for that purpose. Indeed, as Mercator projections were the natural choice for navigators in the Age of Exploration (and gnomonic maps are perhaps the natural choice in the airplane age), many cartographers today think that equal-area maps are the best choice for general use in the modern world.

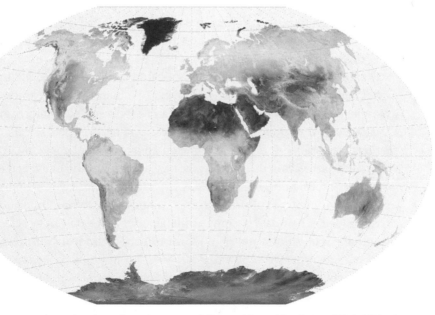

FIGURE 4.13. A modern low-distortion map of the world resulting from a Winkel Tripel projection. NASA image.

Geographic Information Systems (GIS)

Geographic information systems are a product of the computer age, and from the 1970s and 1980s on, they have revolutionized the ancient science and art of mapmaking. Globalization has led to a number of different universal spatial reference systems, analogous to the world clock in time-keeping. Thus, there is the International Terrestrial Reference System and the World Geodetic System. Such systems enable different countries and data sources to merge and exchange data, thus greatly expanding the information that is available for mapmakers way beyond what was possible to present on maps of days gone by. A GIS takes advantage of this plethora of new data, and of the computational power and large memory of digital computers, to capture, store, and manipulate spatial data so that it can be presented in the best manner for different applications.[13]

GIS is used widely in scientific research, resource management and ecology, archaeology, infrastructure assessment and development, demographic studies, disaster management, the spread and control of disease and pollution, and many other fields. An ecologist might use GIS to examine the retreat or advance of vegetation in a Third World drought zone or a city green belt. A town planner might use GIS to assess the infrastructure requirements and population dynamics associated with a proposed large shopping mall development. Wherever layers of information interact, GIS finds an application.

This idea of layering spatial data is central to GIS, and the ability of digital computers to strip off or add layers to a map is a key feature. For example, the land on which a town lies can be represented by topographical data, complete with contour lines, rivers, and perhaps geological information. The same area might be represented by a layer of data containing streets and pathways of different types, perhaps with traffic lights and one-way travel information. Buildings may constitute a third layer, with information about public facilities, stores, hospitals, and so on. Other layers may be added as new data become available. Layers of data can be selected for certain applications and printed on a map, or information from all layers may be presented about a specific location (thus, planning for a new building will require geological and street plan information, but perhaps not contour data).

13. A nontechnical summary of GIS is that of Raju (2004).

As we have seen, an increasingly familiar example of GIS is Google Earth, which allows users to zoom in on a very specific location from, essentially, a very large-scale view of the earth as seen from space. Different information layers are added as the user zooms in to smaller scales. Geographic information systems are likely to increase in both capability and significance as data accumulate and computer technology advances.

CHAPTER FIVE

Early Explorers, Basic Tools

Thus far, we have tackled navigation by reviewing the basic ideas of geodesy and cartography. Now we can turn to the main theme of this book. We will see how these ideas apply to such a wide-ranging (in more ways than one) subject. I begin here with a look at the skills and feats of the navigators and explorers before the European Age of Exploration began in the fifteenth century.

Exploration

If you are anything like me, you are always happy to read about—indeed, are thrilled by—famous explorations from the past. There is something quite appealing about a small group of single-minded individuals who are expanding horizons, using their ingenuity to overcome adversity, and emerging triumphant—sometimes—at the end of a long and arduous expedition. I relate a few of the greatest exploration stories over the next few chapters, partly because exploration and navigation are closely connected, and partly, I must admit, because I find explorers to be hugely impressive individuals whose achievements deserve to be echoed down the centuries.

Exploration is not simply wandering through unknown territory as did, for example, the early hominids that radiated out of Africa into the wider world. These peoples did not need to find their way back, so they did not navigate. True exploration is carried out with a spirit of inquiry (perhaps in addition to more mundane motivations, such as the desire for land or personal wealth) by members of a sedentary civilization who venture far from home and who very much want to return, one day. Hence the link with navigation: they have to find their way home. Some of the later human radiations into previously unoccupied land involved true navigation. For example, the early Polynesians and Micronesians spread out from

Taiwan over 3,000 years ago and did not complete their migrations until the turn of the second millennium, when they reached the last of the unoccupied islands of the Pacific. We will see just how good these seafaring explorers were at seamanship and navigation.

Navigation is all about applying knowledge of the world around us—and above us—to get to a particular destination through unknown or feature-less territory. As noted, I recount the achievements of certain great explorers in this book partly because their stories inspire. If this is not enough for you, if you ask for a closer connection with navigation, if you need a more rigid justification for learning about the achievements of non-navigating explorers, then I suggest that the results of exploration often provided a spur to the development of navigation. Thus, Europeans of the late Middle Ages knew that there was a place somewhere in the east where the fabulous silks and spices that trickled along the Silk Road originated (see fig. 5.1). We will see how this lure arose and how it spurred navigation. Someone had explored these regions, and someone else had brought home the goodies—probably many people, trading the valuable spices hand to hand for thousands of miles. Europeans wanted to find the wealth of the Orient for themselves, to cut out the middleman. Earlier, the ancient Greeks knew that tin came from Cornwall, though they had to explore and discover for themselves the route to that southern peninsula of a far-flung island, in the realm of sea monsters; their trading rivals, the Carthaginians, were not about to tell them. Later exploration led to the serendipitous discovery of the New World, and developments in navigation were necessary to make further transatlantic voyages quicker and safer.

Thus, exploration forms the historical backdrop of navigation and provides the incentive for navigational developments. Many explorers were excellent navigators, but not all: exploration and navigation are two massive subjects with a considerable overlap, but they are not the same. In the remaining chapters I will intersperse the stories of certain explorers with the evolving science of navigation. The explorers will appear roughly in historical order as if, while on a journey, we metaphorically encounter them on the side of the road.

Some of these explorers were not accomplished navigators for the simple reason that they traveled over land and did not need to carefully estimate their position and measure or calculate their direction home; they asked directions or they relied upon the knowledge of those who transported them. Ibn Battuta is an example. You may note many eminent omissions from my list of explorers; this perhaps reflects my Anglocentric

FIGURE 5.1. A faint echo of the Silk Road. This window is in Paderborn Cathedral, in Germany, and dates from the Middle Ages. Yet the theme of three hares—common throughout Western Europe at the time—originated much earlier and much further east: it is Buddhist. The three-hares motif spread westward over the years 600–1500 CE along the Silk Road. Image from Wikimedia Commons, User: Zefram, CC by 2.5.

background and my bias toward navigating explorers. I certainly am not suggesting a lesser achievement for those who are omitted here (Ericson, Zheng He, Gonçalves, Cartier, Cortés, Frobisher, Pinto, Pizarro, Ricci, de Champlain, Quirós, de Houtman, Tasman, Hudson, Bougainville, Mackenzie, Park, Przhevalsky, Lewis and Clark,[1] Franklin, Leichhardt, Livingston, Burton, Peary, Nansen, Amundsen, etc. etc. etc.). They all excelled, but they are not part of our story.

Coastal Piloting

Seafarers since about 3500 BCE have set out to explore the wider world. During the long dawn of navigation, they had very few of the tools that we have seen—no mariner's compass or sextant, for example. They did have sounding poles, or lead lines, to tell them about the depth of the waters they sailed.[2] They built up knowledge of the winds and tides of the seas

1. Lewis and Clark performed many celestial measurements on their expedition, not to determine a travel route but instead for the purposes of mapping the unknown (to Americans) territory they were traversing. A telling indicator of this distinction—between navigational and cartographic observations—is that while Lewis and Clark made their observations, they did not always perform the calculations that would reduce their data to latitude and longitude: they simply retained the data for analysis upon their return.

2. Lead weights attached to knotted ropes indicated the depth of coastal waters. Tallow or wax coating on the lead weight would bring back up sediment from the seabed, which

close to home, and they learned the local currents. Not just directions: useful information could be squeezed out of the temperature and color of water currents, or from the humidity of a wind. From a very early date, these seafarers learned now to steer a course by looking at the night sky or by identifying topographical features of a distant shore. This simple type of navigation is coastal piloting; a sailing ship travels in sight of land. Perhaps in familiar waters the coastal headlands were topped with rock piles to aid seafarers. But even if farther away from home, in new waters previously unknown, the sailor/explorer was still in sight of land and so could return home simply by reversing his course. This is not to say that coastal piloting was confined solely to tentative excursions just a few miles from home port—oh, no, some of the very earliest seafarers wandered much farther than that and deserve to be considered true explorers.

CIRCUMNAVIGATING AFRICA, CLOCKWISE

Some three thousand years ago, the Mediterranean Sea was pretty much owned by the Phoenicians. Phoenicia was not really a political empire in the later sense, but this dynamic group of loosely connected city-states was nevertheless an important power as a trading empire. The Phoenicians were based in the eastern Mediterranean, in modern Lebanon, their principal city being Tyre, which still exists and with the same name.[3] Phoenicians were masters of the sea (though their claim to have invented sailing and fishing is exaggeration), and they plowed the Mediterranean from one end to the other, as merchants carried goods between Phoenician cities to the cities of the ancient Greeks—their main trading rivals—and to barbarian peoples further afield. For example, Phoenicians traded tin obtained from Cornwall, in southwestern England. Extensive trade required written records, and the Phoenician alphabet is considered to be the ancestor of many modern alphabets, including ours.

Overseas expeditions were undertaken by merchant guilds, which were represented in trade negotiations by their king. Trade was the main motivation for the Phoenicians' wanderings, though it is the wanderings themselves that matter more to us.

Trade may not have been in the mind of the pharaoh Necho II, who

---

provided a pilot with more information. There is plenty of evidence that early seafarers, such as Greeks and Carthaginians, used lead lines to navigate bays and harbors.

3. There cannot be many cities still in existence whose heyday was two and a half millennia ago. Tyre possessed a powerful fortress, which was besieged, and the city ruthlessly sacked, by Alexander the Great in 332 BCE.

ruled Egypt from 610 to 595 BCE, when he suggested to his Phoenician allies an exploratory journey to see if Africa was surrounded by water. We do not know his motives, and we do not know the name of the Phoenician captain or the number of his ships or crew. What we do know comes from later writers, principally the Greek historian Herodotus of Halicarnassus.[4] Herodotus tells us, in his *Histories*, that the Phoenicians set out from the Red Sea (familiar waters, being close to home) and proceeded down the east coast of Africa. They traveled for three years, stopping seasonally to plant wheat and staying long enough to harvest it before moving on. The sailors expressed surprise that, when they were at the southern point of Africa, heading due west, the sun was on their starboard side (that is, to the right, or north side). Aided by winds and currents on their southern journey, they were hindered by both once they headed northward, up the west coast of Africa. Eventually they passed through the Pillars of Hercules (the Strait of Gibraltar) into the western Mediterranean. Once there, the Phoenicians were back in familiar waters, and no doubt much relieved, they journeyed across the sea to their home port.

Herodotus discounts some of the reports that he quotes (his sources are lost). Later, when the hugely influential Ptolemy announced that circumnavigation of Africa was impossible, the whole story of the Phoenician circumnavigation was dismissed, though modern historians consider it plausible. Ironically, the very same point that Herodotus disbelieved is the clincher: of course the sun would be to the north of a west-sailing ship in the southern hemisphere. To Herodotus, however, the sun must always be to the south (because it always appeared south of the Greek world). Here was a hint: a northerly sun indicated that world was not flat, but round.[5]

It would be 2,000 years before Africa would next be circumnavigated by man, and the world would be a very different place. Trade would still be a motivation for exploration, but power—and maritime expertise—had shifted to the other end of the Mediterranean. Consequently, the next circumnavigation would be counterclockwise.

4. Halicarnassus is modern Bodrum, in southeastern Turkey, then part of the Greek world. The Greeks would go on to found many cities on the northern side of the Mediterranean, from Turkey to southern France, as we will see. Herodotus wrote a century after the Phoenicians' epic journey, at a time when Greece was on the rise and Phoenicia had given way to its successor, Carthage.

5. The accounts of Herodotus and of other classical authors are widely available. For an annotated translation of Herodotus's *Histories*, see, for example, de Selincourt (2003).

MORE EARLY COASTAL EXPLORATIONS

The glorious exception to the general rule that early maritime explorers kept to the coastal margins, owing to their limited knowledge of navigation, is provided by the South Pacific islanders. The Polynesians and Micronesians made incredible journeys without complex navigational tools; this was possible only because they lived in a swath of the world where winds and currents were fairly predictable, as we will see. Their story merits a separate section: here I am concentrating on European explorers and navigators (and initially, also those from the Near East) because it is from this tradition that we inherit our navigational skills and knowledge. To a certain degree, predictability of winds and currents applied in other parts of the world—for example, in the Indian Ocean, where the annual monsoon could be relied upon to provide southwesterly winds in summer and northeasterly winds in winter. Thus, for example, merchants could, from an early date, make annual voyages between India and East Africa or Southeast Asia. However, weather in the Mediterranean Sea and especially in the North Sea and the Atlantic Ocean was largely unpredictable. To venture far into these waters required navigational tools, which, in the early centuries of recorded history, were few, and so European maritime explorers from those times generally kept close to land.

CARTHAGINIANS, NORTH AND SOUTH

Carthage, near modern Tunis in North Africa, was founded by Phoenicians from Tyre in 814 BCE but had gained independence from its founders by 650 BCE. Indeed, Carthage became the capital city of a large and vigorous empire that lasted 500 years, until it was destroyed utterly by the Romans during the Third Punic War.[6]

Two Carthaginians contributed to the geographical knowledge of the classical world, and both lived in the fifth century BCE. Hanno the Navigator was a Carthaginian king who sailed westward across the Mediterranean Sea with 30,000 colonists and 60 ships, we are told, passing through the Straits of Gibraltar and depositing his passengers at various Carthaginian

6. The word *Punic* comes from the Latin for "Phoenician" and reflects Carthaginian roots. The Carthaginians inherited their ancestors' maritime expertise and their penchant for trade and exploration. The amazing peregrinations of Carthage's most famous son, however, were motivated by neither trade nor exploration: Hannibal led his elephants, and army, across the Alps into Italy during the bloody Second Punic War.

cities on the Atlantic shore of what is now Morocco. Once this task was completed, Hanno kept moving southwestward with a part of his fleet, before turning south and following the African coast around to the Gulf of Guinea and possibly as far as Cameroon. Hanno established trading posts along the way. Upon his return to Carthage, he left an account of his journey; this narrative later was given the name *Periplus* by the Greeks (meaning "a sailing around").[7]

The second of these fifth-century voyagers, Himilco, headed north, once he had passed through the Pillars of Hercules. He followed the coastline of modern Spain and France as far as Brittany before crossing over to southern England and engaging in the tin trade. Upon returning home, he wrote an account of his voyage, now lost, in which he told of a harrowing journey and sea monsters. It may be that Himilco was deliberately playing up the difficulties he encountered to discourage rival Greek traders from venturing out along the Atlantic coastlines and discovering new trading partners. If so, his dissimulation did not work.

Both of these expeditions, and the earlier circumnavigation of Africa, were a testament to the sailing abilities of Mediterranean peoples of the first millennium BCE. No doubt these sailors were also courageous, resolute, and resourceful men, as all explorers must be; however, they need not have been particularly good navigators. They felt their way along coastlines; they got back home by retracing steps. It is true that sailing ships from this period were not as developed as they would become. For example, a medieval caravel could sail a few points into the wind—a feat that was beyond the ships of classical antiquity.[8] But the restriction to coastal sailing was not due solely to the technical limitations of these sailing ships: the mariners of antiquity—in this case, from the long maritime tradition of North Africa and the Middle East—lacked the knowledge of navigation that is required for open-ocean voyages.

7. Hanno's *Periplus* is lost, but his account survives in the writings of many classical authors, including Herodotus, Pliny the Elder (a Roman from the first century CE, who also tells us about Himilco, discussed in the next paragraph), and Arrian (from the second century). Their works are widely available in translation. See also Lacroix (1998). There are brief accounts of the Hanno and Himilco journeys in *Encyclopaedia Britannica*, s.vv. "Hanno" and "Himilco."

8. In the first millennium BCE the most maneuverable ships were war galleys. (Galleys were used for war whereas sailing vessels were used for trade.) Galleys were unsuitable for open ocean travel for fairly obvious reasons. They were open and thus prone to sinking in rough seas. In addition, they required a large crew so could not travel far without revictualing. See Denny (2009) for a technical history of the evolution of sailing ships.

Early Celestial Navigators

The navigational knowledge of ancient Europeans from the classical period also was very limited. Such knowledge as existed had been gleaned over centuries, and came and went with each passing civilization: sometimes knowledge was passed on, and sometimes not. Thus, a very early Mediterranean civilization, the mysterious Minoans of Crete, a Bronze Age people of the third and second millennium BCE, must have known something about navigating open seas because we know that they sailed from Crete to Egypt (fig. 5.2), and in those days such a voyage required spending several nights out of sight of land. The Minoans knew about the Pole Star,[9] and they used Ursa Major (the Great Bear, containing the Big Dipper) to locate it, much as schoolchildren and other skywatchers do today. Such rudimentary knowledge of celestial navigation may have been passed on to the ancient Greeks, or they may have attained it independently. Ursa Major is mentioned in this context in Homer's *Odyssey*. By the third century BCE Ursa Minor was incorporated into the celestial navigator's toolbox, and by the first century CE Beta Draconis (the star Rastoban) was also incorporated.

One of the early Greek explorers was Pytheas, whom we are about to meet. He knew of the Pole Star, and he speculated about the connection between the phases of our moon and the tides. He is known to have used a gnomon, a very early reference to the use of a navigational tool. Another famous Greek mariner was Nearchos, the admiral of Alexander the Great's fleet in the fourth century BCE. Nearchos sailed from the Indus River into the Arabian Sea, across to the Persian Gulf, and up as far as Susa, in the southwest of modern Iran. Eudoxus of Cyzicus, in the second century CE, explored the Arabian Sea from Ptolemaic Egypt, making use of early nautical charts and maps (and no doubt contributing to later ones) that were constructed using stereographic or orthographic projections, thanks to the work of the Greek philosophers we met earlier. Much of this mathematical knowledge would subsequently be lost for over a thousand years.

Pytheas lived in the fourth century BCE in Massilia, the modern Marseilles, on the south coast of France. The city was founded by the ancient Greeks, who spread westward along the northern Mediterranean coast just as their great trading rivals, the Carthaginians, had earlier spread westward along the southern coast. Pytheas is known to us because of an

9. In the time of the Minoans, the Pole Star was not Polaris but Draco.

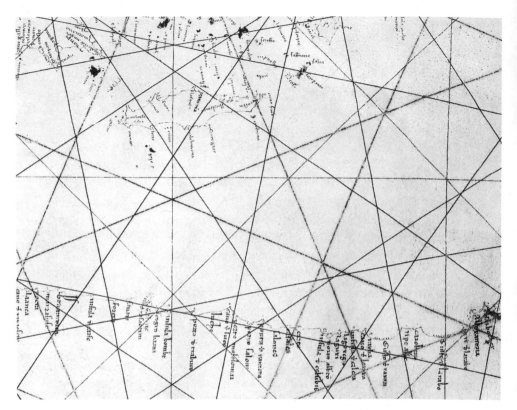

FIGURE 5.2. Detail from a portolan chart of the early 1300s CE. The author of this chart—the earliest cartographic object in the Library of Congress—is unknown but was probably Genoese. This detail shows the island of Crete and the North African shore, including Egypt. Two thousand years before this chart was made, Minoans crossed the sea between Crete and Egypt. Library of Congress, Geography and Map Division, ref.: LC Nautical charts on vellum 3.

extraordinary voyage of exploration that he undertook about 325 BCE to the northern extremes of the habitable world.

Pytheas was primarily a merchant. Indeed, it is quite likely that he made his voyage at public expense to establish trading links with far-flung "barbarians." He probably brought back tin and amber. He claims to have studied the production and processing of tin in Cornwall during his voyage, and he probably succeeded in breaking the Carthaginian monopoly of the Mediterranean tin trade.[10] This perhaps explains the suggestion of

10. Evidence for an established Cornish tin trade with the Greek world is provided by the crude coinage that Britons learned to make (well after Pytheas's visit) mimicking those

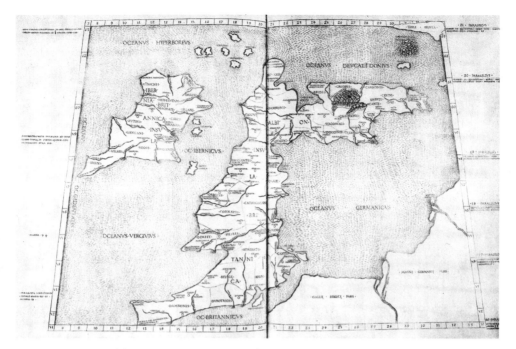

FIGURE 5.3. A fifteenth-century rendition of Ptolemy's map of Britain.

some writers that Pytheas, after setting out westward from his home port, had to dodge a Carthaginian blockade of the Pillars of Hercules (the Strait of Gibraltar).

Once in the Atlantic, Pytheas worked his way up the western coasts of modern Portugal, Spain, and France to Brittany. From there he crossed the English Channel to Cornwall to the tin producers. Proceeding northward, up the west coast of Britain (fig. 5.3), Pytheas observed many details of British life. He noted that "for lack of clear sunshine" Britons threshed their corn in barns, rather than outside as was the practice in Mediterranean regions. He may have described Stonehenge, even in that day an ancient monument. He noted that the longest day in Britain contained "nineteen equinoctial hours"—which is about right for northern Britain. Britons drank a beverage made from grain and honey (mead). He noted the high tides of Britain.

Six days' sailing north of Britain, Pytheas discovered an inhabited "island" he named Thule. This name passed into folklore of the civilized

of the more developed European cultures. The coins displayed Greek features; in particular, they were copies of the Macedonian stater.

southerly regions of Europe as Ultima Thule—the ice-bound northern limit of the world. Speculation has identified the island as Iceland, but Iceland was unpopulated until early medieval times. Many historians now believe that Pytheas was in Norwegian waters (though some have identified Thule with the Estonian island of Saaremaa, in the Baltic). The nights at this latitude, he noted, were of only two or three hours' duration. He traveled still further north, where the sea was frozen. His description is of pancake ice at the edge of drift ice, amid sea, slush, and fog, so common in polar regions. Strabo, a Greek geographer writing some 300 years later, noted that Pytheas spoke "about Thule and about those regions in which there was no longer either land properly so-called, or sea, or air, but a kind of substance concreted from all these elements, resembling a sea-lungs— a thing in which, he says, the earth, the sea, and all the elements are held in suspension; and this is a sort of bond to hold all together, which you can neither walk nor sail upon."[11]

The journey home was probably via the English Channel. Pytheas was a true explorer. We can sense his amazement at what he found at the end of the world. He must have felt very far from home amid that pancake ice. He was knowledgeable about maritime navigation; he knew about Polaris and was familiar with the gnomon, so he could determine latitude by day or night, weather permitting. (At one point in his life he made an estimate of the latitude of Massilia, with an error of only 5 minutes of arc.)

Clearly, there was more to Pytheas's voyage than trade considerations. I get the impression that he was a skilled negotiator, not just because of the trade goods he brought home but also because there is no mention of any hostility from the Celtic and Germanic tribes he encountered. He wrote an influential account of his expedition in a book entitled *On the Ocean*, of which only fragments remain. However, many of the classical writers, including Strabo and Pliny the Elder, must have had access to his work because it is through them that we know about Pytheas of Massilia.

Letting Go of the Shore

Through the centuries, knowledge of navigation accumulated in fits and starts. Knowledge of lead lines was gained and then lost in some parts of the world. They were reinvented in Europe in the thirteenth century for

11. Strabo, *Geographica*, 2.4.1. Translation available online at http://penelope.uchicago .edu/Thayer/E/Roman/Texts/Strabo/2D*.html.

navigating in channels. The Greeks and Phoenicians learned to sail at night and passed on this wisdom. The night sky was hitched to navigation, providing an ever more intricate tool to help a knowledgeable mariner estimate his latitude quite accurately. Such a mariner knew that the Pole Star was not exactly north; it circled the night sky with an angular radius of a couple of degrees, which must be compensated for to achieve accurate navigation. He could say that he was further south than previously because of the height of the noonday sun, or because of the appearance of new star constellations and the disappearance of others. He knew that the sun was lowest in the sky at winter solstice, when it rises and sets further south than at other times of the year. He knew that during equinoxes the sun rose precisely in the east and set precisely in the west.

But from classical antiquity to medieval times, European navigators did not enhance their skills very much. Individually, of course, they learned their craft and improved—or not.[12] New ships were developed, trade fell and recovered following the Dark Ages, but for all the passing centuries, there was little sense of progress in Europe. A maritime navigator from Roman times would still have recognized the techniques used by his counterpart a thousand years later. He might not have recognized the ship on which he stood (an important point, one I will expand upon later), and he certainly would not have recognized the society in which he found himself, but the techniques of navigation were no more advanced than his own— and the world maps he looked at may have been worse than his own, as we have seen.

Ships hugged the shoreline, generally speaking, with a few interesting exceptions. The transition from inshore to open ocean was made boldly by a few people. We have seen that the Minoans mastered the route from Crete to Egypt. Their successors, the Greeks, may have followed: in the Odyssey we read how Odysseus was at sea running before a north wind for five days. We have seen that Phoenicians and Greeks imported tin from Cornwall. It is possible to ply this route without leaving sight of land, but it is a much shorter journey to cut across the Bay of Biscay, though land would fall below the horizon. In the Dark Ages and early medieval times, Vikings took their magnificent open boats across the rough North Sea and across the Norwegian Sea to Iceland (fig. 5.4). Irish seafarers also reached Iceland. Thus, open ocean sailing happened in a few instances, but it was

---

12. As one writer has said, "The better navigators became expert. . . . The poorer navigators just disappeared" (Knights 2001).

FIGURE 5.4. The Oseberg ship, a Viking longship from the ninth century CE. This vessel is on display in the Viking Ship Museum in Norway. Thanks to Waldemar J. Poerner for this image.

not a regular occurrence because of the risks of letting go of the shore and heading into variable and contrary seas without the tools or techniques to accurately establish position or heading.

Yet during this period there were mariners who ventured, as a matter of course, over vast expanses of ocean, and they did so without navigational tools. Instead, they used the accumulated knowledge carefully passed on to succeeding generations. Let us see what they did and how they did it.

South Sea Sailors

The disparate groups of people indigenous to the southern Pacific islands probably originated in Taiwan. From about 3000 BCE they spread from Taiwan throughout the islands of southeast Asia. Two or three of these groups—the Polynesians, the Micronesians, and perhaps the Melanesians —became expert navigators who colonized far-flung and remote islands over a 2,500-year period. From what is now Indonesia and the Philippines they spread eastward to New Guinea and Micronesia, reaching Fiji, and then Tonga and Samoa around 1000 BCE. By 100 CE they had found their

The Great Unknown

It is hardly surprising that, before their Age of Exploration, the European maritime nations clung to the land like a fretful kid with water wings who, learning to swim, clings to the edge of a swimming pool. The terrors of the deep were real (and sailor's superstitions, as well as imagined sea monsters, would not have helped). Consider the photograph here: this is what a mariner in the open ocean sees—if he is lucky and the weather is good. Horizon. Nothing on the surface to indicate direction or distance. If a medieval navigator looked at the horizon from his deck, he might see a distance of 9 km (5 nautical miles). From a crow's nest, the horizon would be twice as far away, at 18 km, so that, on a good day, he might just be able to see the crow's nest of another ship at 36 km.

This distance is *nothing* amid the vastness of the ocean: it would be very easy to get separated from your fleet and find yourself alone and lost. Suppose you were seeking land that rose 1,000 m above sea level. This land would be hidden from your view by the curvature of the earth if you were more than 130 km away—and that distance is less than 1.2° of latitude or longitude. Consequently, European navigators were reluctant to let go of the shore until they were confident that they could reliably estimate (at least one of) these coordinates.

The briny deep. The open ocean provides the navigator of a sailing ship with no surface indication of bearing or of distance traveled. I am grateful to Waldemar J. Poerner for this image.

way a further 2,000 km eastward, to the Marquesas Islands in French Polynesia. This entire region is characterized by low-lying island groups spread thinly across featureless ocean, so that many of these colonizing voyages took place out of sight of land. From the Marquesas, these seafarers (for simplicity, I will call them all South Sea Islanders) boxed the compass: they spread northward, reaching the isolated Hawaiian Islands by about 400 CE.[13] Heading southwest for a couple of thousand kilometers they reached New Zealand between 1250 and 1300 CE. Heading eastward from the Marquesas, they colonized Easter Island (Rapa Nui), probably around 1200 CE. Some argue that they didn't stop there but continued eastward and settled, at least temporarily, on the Pacific coast of South America.

The boats that these South Sea Islanders traveled in were outriggers—open catamarans with a single sail (fig. 5.5). The navigators were specially trained over long periods, and they found their way around their widespread island groups by skillful seamanship and navigation based upon careful observation. None had any navigational tools, apart perhaps from a simply constructed aide-mémoire. How did they do it?

We know quite a lot about the South Sea Islanders' navigation techniques because they are still practiced today, though the knowledge is dying out in the GPS age, and the practice is now perpetuated mainly for cultural reasons. Long voyages were planned far in advance and prepared meticulously. The voyagers would take enough provisions for two months at sea. Their large catamarans could travel 150 miles (about 250 km) per day, and so they were able, in principle, to travel right across the Pacific without stopping for provisions. The navigators, unsurprisingly perhaps, given that many lives depended on them, constituted an elite fraternity. They were taught their craft from childhood: a boy of five or six years would begin to learn the star structures, and he might be put in charge of his first long voyage at age 18 or 20. (Of course, the teachers themselves were experienced navigators, perhaps too old for arduous sea journeys.) The boy had a lot to learn.

The horizon was divided into 32 points, as is our compass. Stars were associated with each point. A novice navigator would learn which stars rose and set at each compass point at different times of the year. Additionally, for insurance (in case of cloud cover) he would learn the stars opposite (for instance, the rising of Antares in the southwest is opposite to

---

13. In fact, archaeological evidence suggests an earlier occupation around 300 BCE, but this first wave either did not survive or was swamped by the later arrivals.

FIGURE 5.5. Kalaniopu'u, King of Owyhee, bringing presents to Captain Cook. This 1781 watercolor is by John Webber, an artist with Cook's expedition. The boats that the South Sea Islanders used on the open ocean were larger versions of such catamarans. Image from Wikipedia.

the setting of Vega in the northeast) and the stars at right angles to these. Thus, a navigator would be able to ascertain his heading from a night sky that was three-quarters covered with cloud.

Elaborate mental exercises from childhood taught a student of navigation the relative positions of islands. He might then be asked for the distance between two of them, or the relative direction, to ensure that the knowledge he acquired was not simply a list of names, but a structure— a searchable database with cross-references. The young navigator would then be taught the locations and directions of sea lanes between the various islands and reefs.[14] Along often-used courses he would memorize local details of reefs, shoals, and seamarks. He knew that whales of this type are often at that location; that these seabirds were here; that such-and-such land-based seabirds could guide him towards that island group. All this data was drummed into the young students. Elaborate visual images were invoked to remember, for example, the star positions ("Parrot Fish," "Jumping Trigger Fish"). Much of the lore was incorporated into chants, which were repeated often until the students knew them by heart.

By day a navigator noted the wind and ocean swell direction and compared these with the previous night's star compass direction (note how this method of estimating bearing assumes constant winds and currents). He recognized different swells and could steer from the alignment of wave peaks that arose when two cross-swells interacted. He knew that there were annual swells from the north, the northeast, and the east that were associated with regular trade winds (as we would call them). The existence of currents was revealed to him by the shapes of waves; he read their direction and strength from patterns of ripples on the water surface. The shapes of clouds on the horizon might help him forecast the weather or indicate the presence of land. The colors of sunrise and sunset also spoke to him about the weather.

Not all boys had the potential to make good navigators, of course, among South Sea Islanders just as among the rest of the world. Clearly, a good memory was essential, along with keen eyesight and acute observational technique.[15]

Poised on the Shore

While Europeans were scuttling around the Mediterranean and South Sea Islanders plied the South Pacific, countless merchants traveled back and forth across the Arabian Sea, the Bay of Bengal, and the South China Sea. Indeed, Indians had been sailing to other lands since the first millennium BCE. On

14. Thus, for example, sailing on the "Sea of Beads" meant traveling between the islands of Woleai and Eauripik on a star course between "Rising of Fishtail" (in Cassiopeia) and "Setting of Two Eyes" (in Scorpio).

15. The fascinating details of South Sea Islander maritime navigation, and their achievements, can be found in Lewis (1994) and Thomas (1987). For an overview of their achievements, see Allen (1980) and Quanchi and Robson (2005). The University of Pennsylvania's Penn Museum maintains an interesting website with much information on the navigators' training.

the brink of the European Age of Exploration, India already had strong naval and merchant fleets plying well-established trade routes to China and Southeast Asia. Following the Arab expansion from the Arabian Peninsula after the birth of Islam, sea travel became an important factor in the Arabs' trade and communication. The new Arab lands contained few large navigable rivers apart from the Tigris and Euphrates, so ocean-going boats were developed, along with the mapping and navigational skills to permit voyaging across the Mediterranean, along the Red Sea, down the east coast of Africa, or across the Arabian Sea (fig. 5.6). By the ninth or tenth century CE Arab sailors possessed a form of compass—almost certainly acquired from further east, as we will soon see—and they navigated with the aid of a *kamal*. This instrument was functionally similar to the cross-staff, though it took a different form. By measuring the altitude of stars it yielded information about latitude and, in combination with the compass and detailed regional maps, allowed Arab traders to travel the open oceans of their part of the world. Throughout the Golden Age of Islam (roughly the seventh to the thirteenth centuries) the Arabs contributed much to navigation by integrating geographical and technical knowledge from a wide area, from sources that were previously independent of one another.

The compass originated somewhere in the East, although the date and place are uncertain. The Chinese first applied naturally occurring magnetic material to navigation, probably in the eleventh century CE, but the knowledge of such materials appears to be much older. Magnetite is a mineral of iron oxide that was called *lodestone*. Initially, lodestones seem to have been used in China for *feng shui*. Subsequently, they were used to magnetize small needles of iron or steel which, when free to turn, orient along the earth's magnetic field lines. Navigators found that when magnetized needles were floated on water (in a wisp of straw or on a small piece of wood) so that they were free to turn, they could orient north-south.

Wherever and whenever the application of magnetic material first occurred (there are claims that compasses existed in fifth-century India), the use spread westward. By about 1300 CE the Europeans had learned of compasses and had improved them. The dry compass—a magnetized needle free to turn on a pivot (fig. 5.7)—is a European invention, made sometime before 1410 CE.[16] The mariner's form of this compass was steadily improved until it became a widely used, though not fully trusted, instru-

16. The idea of placing a magnetized needle on a pivot may have been Chinese, but the idea of placing the instrument on a gimbal was certainly European.

(a)

(b)

FIGURE 5.6. Arab dhows. These examples are coastal vessels, but larger versions carried goods and people across open seas. Photo by Matt Crypto; sketch of a Maldives boat by Xavier Romero-Frias.

ment of navigation. Mariners knew that a compass needle pointed approximately north, but how close to true north appeared to be variable, depending upon location. The cause—magnetic variation or declination—was not then understood, so navigators of the fourteenth and fifteenth centuries regarded the compass as useful but unreliable.

On the brink of expanding into the world, Europeans had little in the way of equipment to take them there. Indeed, it was finding the right equipment that enabled them to explore and expand. The only strong indicator that something might be stirring in this part of the world, to my mind, was the portolan charts (fig. 5.8). These charts, recall, represented accumulated practical knowledge of the waters and shores of the European world, with information about harbors, currents, and wind patterns, and every bit of useful knowledge that could be accrued and recorded. The directions indicated on these charts were only approximate because they were obtained by use of compasses, but magnetic variation was not a problem; the navigators who used these charts also used compasses, so the common declination error canceled. Only over time would the drifting geomagnetic field cause differences between the directions indicated on portolan charts and the same direction displayed on a mariner's compass; thus, the charts required occasional updating.

Why do I consider the portolan charts to be an indicator of things to come? These charts suggested that Europeans were once again learning how to learn; they were applying knowledge to improve things, to enable them to do something (in this case, piloting and navigation) better than before. The habits of intelligent observation and applied learning had largely been lost since the end of the classical Greek era, over a thousand years

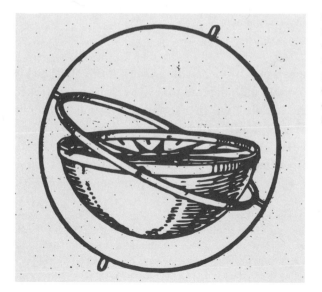

FIGURE 5.7. A mariner's compass, cradled in a gimbal to reduce the effects of a pitching and rolling deck, shown here in a drawing from 1570. Image from Wikipedia.

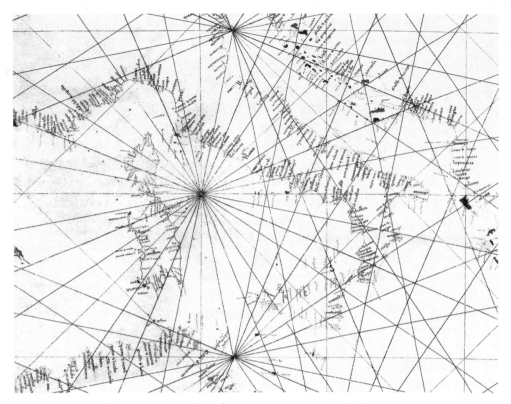

FIGURE 5.8. Part of a fourteenth-century portolan chart, showing Italy and surrounding areas. This detail is from the same chart as that of figure 5.2. The radiating lines are rhumb lines, for the benefit of mariners.

earlier. Crucially, European ships were also improving at this time, owing to the convergence of two ship-building traditions. The Crusades had taken Christian soldiers to the Middle East, in a series of invasions of Islamic lands that still rankle there today. However, these invasions also brought northern European clinker-built ships, with their newly invented rudders, into Mediterranean waters, where the European sailors encountered southern European carvel-built ships.[17] Over the next couple of centuries,

17. *Clinker* (also known as *lapstrake*) and *carvel* refer to the method of constructing the outer hull of a wooden sailing ship. Clinker-built ships had overlapping planks, as with the famous Viking longboats. Such boats were strong—and beautiful—but were limited in size by the plank length. Carvel-built ships had the planks abutting, like siding on the wall of a house, and were not restricted in size to the length of a plank; planks could be abutted end-

the best features of each tradition—southern clinker hulls and northern rudders—were combined to produce tough ocean-going vessels that were rigged to sail a few points into the wind. These ships eventually became much larger, culminating in the magnificent sailing vessels of the Age of Sail in the early nineteenth century. At the time of European expansion, they were still small—but seaworthy.

The last part of this chapter answers a question that may have occurred to you. Why did the Europeans want to explore the rest of the world? There are several reasons, including a perceived need to convert the peoples of the rest of the world to Christianity, but the main reason was trade. The Orient in particular represented fabulous wealth compared with backward little Europe. The Portuguese and Spaniards (at first—and then the English, French, and Dutch) set out to acquire some of that wealth for themselves, initially by trading for it and later by conquest. But, you rightly ask, if they and everyone else lacked the means to sail around the world, how did Europeans know about the wealth of the Orient? The answer lies in the overland trade routes of the justly famous Silk Road that brought merchants and goods from the East to the West (fig. 5.9). To see how this Silk Road emerged, we must back up 1,500 years and look to China.

Zhang Qian and the Silk Road

The Han dynasty was the first that unified large tracts of China. It reached its peak during the first centuries CE, at the same time as three other large empires. The Kushan Empire was a loose confederation of horsemen and traders who occupied vast tracts of land to the west of China, from modern-day Uzbekistan to northern India. The Parthians were further west, in modern Iran; they were famed for their horse archers. To the west of Parthia lay its enemy, the Roman Empire. Thus, for a few centuries, there was a region of the Old World with very few national borders. A traveler could wander overland from the shores on the North Pacific, in eastern China, to the shores of the North Atlantic, in southern Scotland, and in doing so cross only three borders. Such a unification of peoples had not been seen before and has not happened since. This lack of borders encour-

---

to-end. See Denny (2009) for medieval sailing ship development. For the wider state of European preparedness and technological know-how on the eve of expansion, see Bedini (1998).

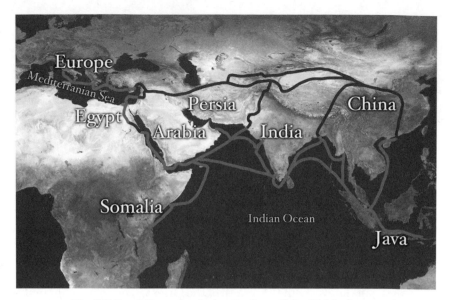

FIGURE 5.9. The Silk Road. From classical antiquity until the Middle Ages, traders used this network of routes to transport high-value goods from the Orient to the Occident. Goods from China and India traveled over land; spices from the Spice Islands near Java came to the West by sea. Image by San Roze.

aged the development of long-distance land routes for trading. The Silk Road emerged at about this time, around 100 BCE, and persisted in one form or another for many centuries, long after all four of the empires of classical antiquity had fallen apart.

Zhang Qian[18] witnessed the beginnings of the Silk Road. He was an envoy sent westward by the Han emperor Wudi, with an escort of 100 men, to seek out allies against the marauding Xiong-nu "barbarians." However, he was captured by the Xiong-nu and held prisoner for 12 years. Despite being "barbarians," the Xiong-nu treated him well. Eventually, Zhang Qian escaped with his men and fled further west, still trying to fulfill the mission of his emperor. He was seeking the Yueh-chi, a nomadic people who, he had heard, might be willing to fight against the Xiong-nu and who had magnificent horses, better than those of the Chinese. During his period of captivity, the Yueh-chi had moved even further to the west, and Zhang Qian followed them. He traveled across the Tian Shan Mountains as far as Bactria, where he found the Yueh-chi and their fabulous "heavenly horses."

He stayed a year with these people but could not persuade them to wage

18. Pronounced "Jang Chyen."

war against the Xiong-nu. The Yueh-chi were very wealthy and at peace. Both of these observations surprised and impressed Zhang Qian, who noted their jade (much valued in China). They noticed his silk and agreed to trade with China. The Yueh-chi, from their base in Bactria, were also trading with India (to the south) and with peoples further westward who were unknown to Zhang Qian. His hosts informed him that 200 years previously, a great conqueror from the West (Alexander the Great) had overrun Bactria. Zhang Qian noticed the Greek coins, script, and sculptures that had accrued from this Western contact. These, and the presence of Chinese bamboo and cloth, showed Zhang Qian the extent of Yueh-chi trade.

He left the Yueh-chi and passed through the territory of his enemies once again on the way home. Again he was captured—this time he was held for a year—and again he escaped. After 13 years away from home, Zhang Qian returned to China in 125 BCE, with a Xiong-nu wife, a son, and one companion. He was the first man to bring back a reliable account of Central Asia to the Chinese court, and for this service he was rewarded by appointment to a high office within the imperial bureaucracy. His emperor wanted the "heavenly horses" and so established firm trade links with the Yueh-chi, thus joining China with the emerging Silk Road to the west.[19]

Ibn Battuta

For a different view of the Silk Road—from the other end and from a time much closer to the era of European expansion—consider the amazing travels of Ibn Battuta. His story also serves to show why the overland route to the Orient was closed to medieval Europeans.

In 1325, a 21-year-old Moroccan called Abu Abdallah ibn Battuta set out on a journey from Tangier that was longer than his name.[20] Initially, he was traveling to Mecca on pilgrimage, the *hajj* that all Muslims should try to do at least once in their lives. He joined a caravan and crossed the Maghrib to Alexandria in Egypt—a 3,500-km first leg of an epic journey that would keep the young legal scholar away from home for the next 24 years.

19. To learn more about Zhang Qian's exploration and about the Silk Road in general, see Gosch and Stearns (2008) and Wood (2002); also *Encyclopaedia Britannica*, s.v. "Zhang-Qian."

20. This is the shortened version of his name, his full handle being Shams al-Din Abu Abdallah Muhammad ibn Abdallah ibn Muhammad ibn Ibrahim ibn Muhammad ibn Ibrahim ibn Yusuf al-Lawati al-Tanji Ibn Battuta.

From Alexandria, Ibn Battuta traveled to Cairo and Damascus before making the 1,500-km journey to the holy cities of Medina and Mecca. At this point he could have returned to his home city of Fes, but instead turned east and traveled through the land that is now Iraq and southern Persia before crossing the Persian Gulf to Yemen, on the Arabian Peninsula. During these early years of his travels, he spent most of his time on the move—seemingly from a sense of adventure. He kept going southward, down the East African coast as far as Kilwa in modern Tanzania, then a Muslim sultanate growing prosperous on gold and slaves. From Kilwa he returned to Mecca (this time staying for a year) before passing through Syria to Anatolia, birthplace of the Ottoman Empire, during the period of the initial Ottoman expansion.

Ibn Battuta crossed the Black Sea from the Anatolian port of Sinope to the Crimea, then part of the Golden Horde. This massive empire stretched across the southern Russian steppe and was a branch of the larger Mongol domains that had converted to Islam. He journeyed with the khan as far as Astrakhan, on the Volga, before making a side trip to the then-Christian city of Constantinople (modern Istanbul). He returned to Astrakhan and from there headed further east, past the Caspian Sea to Bukhara and Samarkand, before turning south through the Hindu Kush and settling in Delhi, India, where he stayed for several years. At this time Delhi was a Muslim sultanate (the Tughluq dynasty), and Ibn Battuta used his knowledge of Islamic law to talk himself into a job with the sultan. He stayed in India for the next 12 years, though he made many side trips—to Ceylon, Sumatra, Vietnam, and to Yuan China—as the Tughluq sultan's ambassador. Finally, after nearly a quarter-century away, in 1349 he returned home via Hormuz and Damascus.

But Ibn Battuta was not yet done traveling. Three years after his return to Morocco, he was off on another journey. This time he headed south through the Sahara to the nation of Mali and its eastern neighbor, the Songay kingdom. His detailed descriptions of these states are the best source of information about them during this period for modern historians of sub-Saharan Africa. He was back home after two years, and this time he stayed there. He dictated from memory the story of his 25-year journey in a book, *Rihla*, most of which is accepted as true. This account did not become well-known to scholars—even in the Muslim world—until in the nineteenth century. Ibn Battuta was well aware of the scope of his travels. In *Rihla*, he comments that he met someone named Abdullah al-Misri, "the traveler, and a man of saintly life. He journeyed through the earth, but he never went into China nor the island of Ceylon, nor the Maghrib, nor al-Andalus, nor the Negrolands, so that I have outdone him by visiting these regions."[21]

Ibn Battuta had survived an attack by Hindu rebels, civil wars and the Black Death (in Palestine during his homeward journey), and the rigors of fourteenth-century travel, often through arid and remote tracts of land. Yet, in some ways his travels, a third of the way round the world and back again, were easier than they would have been for a Christian from Europe. This is because the Muslim world was large and spreading during the fourteenth century. Christian travel through Muslim lands was often proscribed, despite Marco Polo's good fortune, as we will see. It was partly because of this difficulty—in particular, for Christians traveling along the Silk Road—that led European maritime nations to develop ocean-going ships so that they could trade with the fabulously wealthy (to them) Orient by sea. Fifteenth-century Europe knew that great wealth lay further east, in regions they could not reach by land. By this time they had the means, as well as the motive, to get there by sea.

21. Dunn (1989). Ibn Battuta is now well-known online. For a more detailed account of his travels, see Dunn.

CHAPTER SIX

Europe Discovers the World

The rapid exploration of the world's oceans and the beginning of European overseas empires were accompanied by gradual improvements in navigation. Here, we look at the developments of the fifteenth and sixteenth centuries, and I show how some of the early navigation instruments were used.

The Search for Spices

For fourteen hundred years, the fabulous wealth of the East—everything from silk to steel and spices—had been carried by mule, camel, or wagon along the Silk Road to the humbler civilizations in the West. Rather than being a single physical track, the Silk Road was instead a broad route that terminated in the Near East, though some of the goods kept moving westward to adorn the households of the wealthy. Among the most sought-after items by medieval Europeans were spices: cloves, nutmeg, peppers, cinnamon, saffron, mace, cumin. The first three were particularly desired for their properties of flavoring and preserving meat, and Venetian merchants shipped them out of Constantinople (modern Istanbul) to Western Europe by the ton. Most of these spices originated in the Moluccas, an island group in modern Indonesia (fig. 6.1). Other spices that made their way along the Silk Road included aloewood, camphor, ginger, sandalwood, and turmeric from China and cardamom, pepper, and sesame from India. (The word *spice* in medieval times had a broader meaning in many European languages; it included products that today we would classify as perfumes and drugs.) The long journey of these spices, passing from hand to hand, made them very expensive.

In the previous chapter we saw how the Silk Road arose. Here, we concentrate on the European effort to circumvent it. The Silk Road passed through mostly Muslim lands that were usually closed to Christian travelers. An exceptional window opened to the East with the Mongol conquests of much of Asia; during this time Marco Polo, his father, and his uncle traveled the Silk Road to China. Polo left Venice in 1271, at the age of 17, and returned 24 years later with a story that would astonish the whole of Europe. He learned Tatar languages; he met Kublai Khan, who appointed him an ambassador to outlying regions; he saw the wealth and extent of China—cities, palaces, bridges, canals, and ocean-going junks. His story was nearly never told: in a Genoese jail after returning home, he was persuaded by a fellow prisoner named Rusticiano or Rustichello, himself a writer, to publish a book. *Travels* was a great success, but at first many readers did not believe what they read or assumed that Polo was exaggerating. On his deathbed in 1324 Polo said, "I have not told one half of what I saw." His book became very influential; for example, portolan charts and an important map of 1459 incorporated features from his *Travels*.[1]

Kublai Khan's death closed the Silk Road for a time. Later, that road was barred to Europeans with the fall of Christian Constantinople to the burgeoning Ottoman Empire. However, the genie had been let out of the bottle. Europeans knew that China and India were fabulously wealthy and that spices came from somewhere over there. Because they could not reach the Spice Islands or any of the Eastern trading ports by land, the maritime nations of Western Europe made determined efforts over the next 300 years to reach them by sea. The spiceries, as the Moluccan Islands came to be called, provided the initial stimulus for all the subsequent exploration in the fifteenth, sixteenth, and seventeenth centuries. All the expeditions and explorers that you have heard of from those days, and all their extraordinary efforts, were aimed at reaching the spiceries, trading with the local rulers for exclusive deals, and bringing riches in the form of spices back to Europe.[2]

Columbus, da Gama, Dias, Magellan, Cartier, Hudson, Barents—all

1. The 1459 map is the Fra Mauro map, now on display at the Biblioteca Nazionale Marciana in Venice. Interestingly, another pre-Columbus map, the Pizzigano Chart of 1424, portrays a reasonably accurate delineation of Europe, with the odd addition of two substantial (and entirely fictitious) islands out in the Atlantic named Antilia and Satanazes. Some historians think that these islands represent the findings of mariners who had earlier crossed the Atlantic and returned with news of distant lands.

2. For details on the European search for the Spice Islands, and for the Age of Exploration, see Bedini (1998); Bergreen (2007); Boorstin (1983); *Encyclopaedia Britannica*, s.vv. "Vasco da Gama," "Ferdinand Magellan"; Brown (2003); *Encarta Encyclopedia*, s.v. "Magellan"; Haase and Reinhold (1993); Milton (1999); Owen (1979); and Ravenstein (1998).

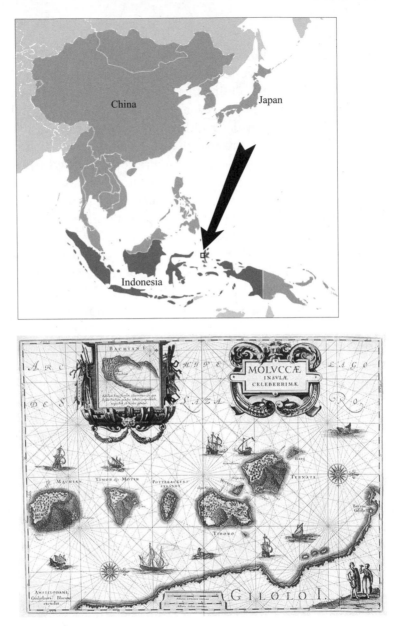

(a)

(b)

FIGURE 6.1. The Spice Islands. (a) Location of the northern Moluccas (small rectangle), which in historical times were the only sources of spices such as nutmeg and cloves. (b) A 1630 map of the Spice Islands by Willem Blaeu (note that north is to the right); this map is roughly the area shown in the rectangle of (a). The first maps of this region are from a century earlier; by the time of this seventeenth-century map, the Spice Islands were controlled by the Dutch. (a) Modern map adapted from a map courtesy of the CIA; (b) 1630 map from *Atlas Novus*, by Willem Blaeu.

these famous names, and all of their achievements that make up such a significant part of history, were motivated by the lure of the Spice Islands. Columbus was not trying to find America; he was looking for the Indies, where the spices came from. Magellan, rounding Cape Horn, was not trying to circumnavigate the world; he was searching for spices. The search for the Northeast Passage over the top of Asia was a search for a quick route to the spiceries. The first attempts at a Northwest Passage across the top of Canada were to find a quicker route to the spiceries. European ships, once they had found their way, buzzed around the Moluccan archipelago likes wasps around a jam jar. For three centuries, the spice of life in Europe was, well, spice (fig. 6.2).

Prince Henry the Navigator set up a "school for exploration" at Sagres, in Portugal. Here, mariners from around the Mediterranean world learned navigation, improved their vessels, and developed ocean-going ships (the *caravel* in particular; see fig. 6.3). Henry sought to systematize the study of navigation and so improve upon current practices. Under his direction, fleets of small Portuguese ships felt their way down the west coast of Africa and into the Atlantic. The verdant and unpopulated island of Madeira was

FIGURE 6.2. Spices! Photo courtesy of heydrienne.

FIGURE 6.3. Replicas of the three small ships that crossed the Atlantic on the first voyage of Christopher Columbus. These replicas also made the journey from Spain to the Chicago World's Fair of 1893. From left to right these are the *Pinta*, the *Santa Maria*, and the *Niña*. The *Santa Maria* was a carrack, while the *Pinta* and the *Niña* were caravels. E. Benjamin Andrews, *History of the United States*, vol. 5 (New York: Scribner's, 1912).

discovered in 1420,[3] and the Azores—a third of the way across the Atlantic —in 1432. The obstacle posed by Cape Bojador, a headland 150 miles south of the Canaries on the African coast that had dangerous shoals and contrary winds, was finally overcome in 1435 by the fifteenth fleet sent out for that purpose—and a year later, slavery out of western Africa began. (In fact, the barrier posed by Cape Bojador seems largely to have been psychological. It inspired terror in the minds of superstitious sailors, who believed ancient legends about the unnavigability of waters on the other side.)

The main goal, though, was to find a way around Africa to the spiceries. The influence of Ptolemy was waning in some quarters,[4] and Henry

3. In fact, Madeira appears on an Italian portolan map of 1351, but it seems that such discoveries did not count without royal patronage: witness the "discovery" of North America by Columbus, several hundred years after the Viking settlement in Newfoundland was first constructed. It is also likely that fishermen from Western Europe—Bretons, Basques, Cornish, and Irish—found the Grand Banks and the Newfoundland coast before Columbus set sail.

4. For instance, Nicolo de Conti, a Venetian merchant, had spent 25 years traveling the west coast of India, Ceylon (Sri Lanka), Sumatra, Burma, and Java before returning to

thought that sailing around Africa might be possible. By the time of his death in 1460, the Portuguese had reached Sierra Leone, within 10° of the equator, a feat not achieved since the Carthaginians. It was Bartolomeu Dias who rounded the Cape of Good Hope, satisfied himself that this was the southernmost point of the continent, and headed back home with the welcome news in 1488.

The Spaniards did not sit idly by watching their neighbors strive to reach the Spice Islands. An Italian adventurer was on the docks to greet Dias upon his return. The Italian had tried to convince half the royal houses of Europe to sponsor him on a westward expedition. Christopher Columbus believed Ptolemy's underestimate of the size of the world and thought it would be easier heading west to find the spiceries than sailing eastward. In 1492 King Ferdinand and Queen Isabella of a newly unified Spain believed Columbus enough to pay for his first expedition (figs. 6.4–6.5). Columbus found landfall (probably at Grand Turk in the southern Bahamas), explored some of the Caribbean islands, and brought back the welcome news that he had found the Indies. He thought that Cuba was the island of Cipango (probably Japan) at the eastern extremity of Asia—a misconception that he took to his grave.

Following Columbus's discovery of the New World, a papal bull allocated all lands to the west and south of Spain to the Spanish king. As a result, the Portuguese had to redouble their efforts to find an eastern route to the spiceries. In 1497 the enterprising and cruel Vasco da Gama sailed south with a fleet of four small ships and 150 men. Accompanied on the first leg of his voyage by Dias, he passed the Cape Verde Islands and swung out to sea at Sierra Leone to avoid the Guinea current and the doldrums. (Most fifteenth-century sailing involved creeping along coastlines, but Columbus and da Gama took long leaps of faith—a sign of things to come.) He rejoined the African coast farther south, rounded the Cape, and fighting scurvy, storms, and contrary currents, revictualed in Mozambique. At the port of Malindi (in modern Kenya) he took on board a pilot and crossed the Indian Ocean, reaching Calicut, the most important trading post in southern India, in May 1498. He was back home six months later, with news of an eastern route to the wealth of the Orient.

Da Gama left a trail of dead bodies in his wake—there were many more after his second trip—and began two centuries of Portuguese prosperity

---

Europe in 1444. He speculated publicly about the possibility of reaching India by sailing south of Africa.

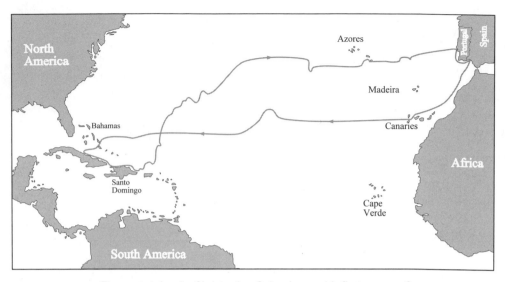

FIGURE 6.4. The route taken by Christopher Columbus on his first voyage. On the outward journey, in the mid-Atlantic, he noticed that his compass no longer pointed at Polaris—the first recorded observation of magnetic declination. Earlier German maps suggest that the phenomenon of magnetic declination was known, but this knowledge was not explicitly recorded.

FIGURE 6.5. Christopher Columbus. The first voyage of Columbus is celebrated on both sides of the Atlantic—and in the middle of it. This statue is on the island of Madeira. Photo by Waldemar J. Poerner.

based on trade with the Orient (though the Venetians were less than pleased that their Constantinople trade in Silk Road goodies would henceforth be undercut). Six months later Pedro Cabral, following da Gama, accidentally discovered Brazil on his outward journey, leading to the creation of a Portuguese empire in the New World.

The Spanish throne backed another Portuguese captain, Ferdinand Magellan, to find a western route to the spiceries, having lost confidence in the claims of Columbus. By this time, Amerigo Vespucci had explored part of the South American coast and had returned to Europe convinced that it was not Asia, but a new continent.[5] Magellan's journey began badly and went downhill from there. Working his way down the coast of South America, trying to find a way through the continent to the Pacific (as he would later name that ocean) like a fly repeatedly banging into a window, he was obliged to execute mutineers. Passing through the strait that now bears his name, he and his crew endured a horrible 98-day journey across the Pacific Ocean without replenishing food or water before they eventually reached landfall at Guam. In the Philippines Magellan was killed after taking sides in a local war. His deputy del Cano (or Elcano) completed the first circumnavigation of the world in September 1522 (figs. 6.6–6.7).[6] Despite losing four out of five ships and all but 18 of the original 270 crew members, this expedition made a profit by virtue of the 26 tons of cloves and cinnamon it brought back home.[7]

Thus, empires were carved out as a by-product of the search for spices. The English, just beginning their rise to maritime dominance, were not to be left out of the spice trade, but because the Spanish and the Portuguese had closed the southern routes, the English tried to get to the Spice Islands via the northeast and then the northwest, establishing trading posts and

5. A German geographer, Martin Waldseemüller, wrote in 1507 that "since another fourth part [of the world] has been discovered by Americus Vesputius, I do not see why anyone should object to its being called after Americus the discoverer, a man of natural wisdom, Land of Americus or America, since both Europe and Asia have derived their names from women" (Haase and Reinhold 1993, p. 122). Waldseemüller would produce a world map in 1513 that was the first to cover 360° of longitude.

6. Magellan is generally given credit as the first person to circumnavigate the world because of the sum total of his earlier travels and his final journey; del Cano was the first to circumnavigate during a single voyage.

7. One of da Gama's party later wrote about seeing four Moorish vessels in a port that were "laden with gold, silver, cloves, pepper, ginger and silver rings, as also quantities of pearls, jewels and rubies, all of which articles are used by the people of this country" (Ravenstein 1998, p. 23; also quoted in Jardine 1996, p. 80). Note how the spices appear on a par with precious metals.

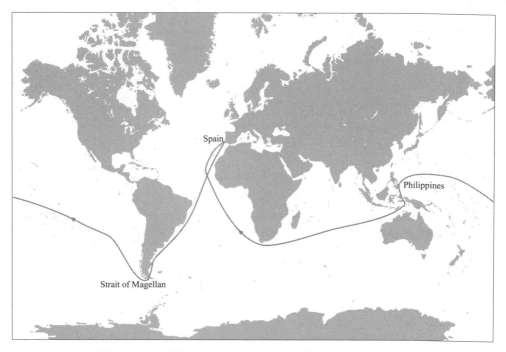

FIGURE 6.6. The voyage of *Victoria*, 1519–22. Initially, Magellan was in command, but after his death in the Philippines, the ship was brought home to Spain by del Cano.

colonies in Canada. Eventually, it would be the Dutch who won the Spice Islands (and set up their main overseas colony in Batavia, in Indonesia, to oversee the spice trade). Tough and capable ships manned by tough and capable crews had crossed the world and, in doing so, changed it.[8]

Standing Off and Shooting the Sun

If the Portuguese dominated the first phase of European exploration, the Spanish dominated the second. They were the first Europeans to establish an empire in the New World, and within that empire they discovered tons of gold and a mountain of silver. All this wealth necessitated increased

8. The importance of the Spice Islands, as compared with the vastly larger New World discoveries, is brought home by considering the Treaty of Breda, which ended a seventeenth-century Anglo-Dutch war. The English gave up Run, a tiny island in the spiceries, to the Dutch in return for the island of Manhattan. Both parties at the time thought that the Dutch got the better deal.

FIGURE 6.7. Detail from a sixteenth-century townscape of Seville, home port for the Spanish expeditions of discovery, by Alonso Sánchez Coello. By royal decree, all ships sailing between Europe and the New World had to pass through the port of Seville. Image from Wikipedia.

travel across the Atlantic Ocean. The search for spices continued with exploration of the Indian and Pacific Oceans by Spain (and Portugal; England, Holland, and France joined later). This search led to the Spanish colonization of the Philippines, after Andrés de Urdaneta had found the Spice Islands and discovered a new route back home, later known as "Urdaneta's route." This route took advantage of the *volta do mar*, as his Portuguese competitors called the regular ocean-current circulation patterns.[9]

The two main convoy routes adopted by the Spaniards to ship home

9. There are five major circulation patterns, or *gyres*: in the north and south Atlantic and Pacific Oceans, and in the Indian Ocean. The northern hemisphere gyres circulate in a clockwise direction, and the southern gyres circulate counterclockwise.

Columbus as Navigator

Navigational tool were improving only gradually during the Age of Exploration, as we have seen, and the practices of navigators during this period reflect the mixture of old and new. Consider the crude manner in which Portuguese ships navigated their *volta do mar* (return journey) to home port after exploring the west coast of Africa. The winds and the Guinea Current obliged them to adopt a broad sweep out to the open ocean on the way home, a route known to them as the *Elmina track*. They found themselves heading northward, out of sight of land, needing to turn eastward in order to reach Portugal. Their navigational method was as simple as it was inefficient: they headed up the Elmina track until the Pole Star was at the same altitude as it was at Lisbon, and then they turned east. Given the long distance they still had to travel, this was a rough-and-ready method at best because winds and ocean currents could easily drag them off course. On the other hand, they knew to compensate for the orbit of Polaris (by noting the orientation of the nearby *guard stars*), so at least that source of error was accounted for. Technically, this mode of navigating is known as *altura* (height) sailing—a less precise version of latitude navigation, which followed on from it.

A detailed analysis of his writings shows that Columbus, though a Genoese in the employ of the Spanish court, was at heart a Portuguese navigator. His training and experience, and his use of the tools of his trade, all point to this conclusion. Columbus had an astrolabe and quadrant, with which he measured the altitude of the sun and stars. He had crude charts, a lead line, a sandglass, and possibly a sundial. Of these instruments, only the quadrant and astrolabe would have been unfamiliar to a mariner three centuries earlier.*

One aspect of sailing that had not changed much since earlier times—and would not for centuries after Columbus—was the superstition of sailors. This characteristic of his crew (or perhaps their more pragmatic worries) caused Columbus on his first outward voyage across the Atlantic to hide from them a puzzling deflection of his compass away from the Pole Star and the true distance that his small flotilla had covered.

* Bedini (1998, pp. 512–14) provides details about Columbus's navigational training and skills.

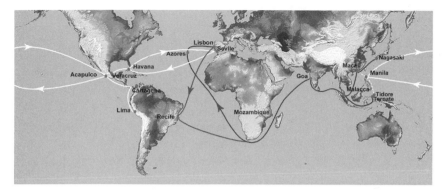

FIGURE 6.8. Sixteenth-century maritime trade routes of the Spanish (white lines) and the Portuguese (dark lines). World travel was increasing, and with this increase came a need for improved navigation. NASA topographical image, modified by Uxbona to show maritime routes.

gold, silver, spices, and other high-value commodities from their colonies are shown in figure 6.8. The richness of these Spanish "treasure fleets" attracted unwanted attention from buccaneers (licensed pirates from other nations, and in particular, the English, Dutch, and French). The Spanish and Portuguese got together and carved up the world between them in a couple of treaties (fig. 6.9)—a display of confidence (as well as arrogance) in their new-found abilities to travel and project power all over the globe.

The speed with which the Spanish, in particular, established and exploited their overseas colonies is quite amazing. To provide just one example, within a half century of Columbus's "discovery" of the New World, the Spaniards had found huge deposits of silver ore in a mountain in remote Bolivia (then part of Peru). The mountain soon became honeycombed with mines and rapidly gave rise to the town, and later city, of Potosi in the bleak highlands (at an altitude of 4,000 m, or 13,000 ft). Thousands of indigenous people, plus slaves from Africa, died laboring to gouge the silver ore out of *Cerro Rico*, the "rich mountain" (fig. 6.10). This ore was then carried by pack animals—llama and mules—down to the coast. Thousands of tons of silver ore generated billions, some say trillions, of dollars (in today's money) for the Spanish treasury over the next two centuries. So much wealth poured in that the economies of both the New World colonies and the home country were profoundly affected (in the long run, not always for the better).[10]

10. For a detailed account of the economic impact of New World silver production in the colonial period, see Garner (1988).

FIGURE 6.9. Sixteenth-century Spanish and Portuguese territorial claims. The first European nations to explore the world arranged between themselves how to carve it up. The Treaty of Tordesillas of 1494 gave to Spain everything west of the W46 meridian; the Treaty of Zaragoza in 1529 gave Spain everything east of the E142 meridian—Portugal got the rest. Of course, this is an oversimplification; these treaties referred only to non-Christian parts of the world that these two countries considered to be up for grabs.

To aid such seaborne travel, be it exploration of unknown regions or the trade of treasure fleets returning home laden with wealth, navigation was improved both in terms of techniques and accuracy. The navigational tools available to a mariner during this period were a cross-staff and an astrolabe for celestial measurements, an hourglass or a water clock for timekeeping, a globe and calipers, crude charts, a mariner's compass (at this stage not yet mounted on gimbals), and dead reckoning. Empirical knowledge of the movement of heavenly bodies had advanced to a level that permitted pre-diction of the time of lunar eclipses. By measuring the local time of a lunar eclipse at some distant location of unknown longitude and comparing that with the known time of the eclipse back in the home port, navigators could calculate longitude. (Later we will examine in detail the connection be-tween time and longitude.) By the end of our period, in the late sixteenth century, the "lunar distance" method of estimating longitude was known in principle, though data were insufficiently accurate for it to be used in practice. This idea had been put forward in the late 1400s by a German mathematician and astronomer who wrote under the name of Regiomon-tanus (his real name was Johann Müller). Then in 1533, Gemma Frisius

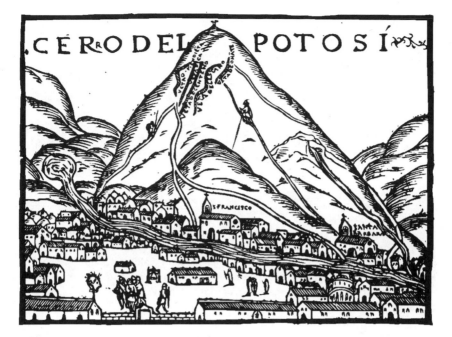

FIGURE 6.10. An early (1553) view of the Cerra Rico silver mines in Potosi, in modern Bolivia. This mine was opened in 1545 and generated huge wealth for Spain, which the "treasure fleet" brought back home. A. Skromnitsky.

suggested that longitude could be established by measuring time differences in different parts of the globe.[11]

The ideas were there, and people were beginning to get a grip on the problems of navigation scientifically—mathematically—for the first time since the ancient Greeks. But the world was not yet ready to accurately estimate longitude: that was a hard nut to crack, and a practical solution was not achieved until the eighteenth century. The best that could be done in the sixteenth and seventeenth centuries was to "guesstimate" longitude by *dead reckoning*, a method for estimating distance traveled. This important technique is unpacked in the next section; here, I discuss simpler and older methods of navigation based on angle estimation.

At the end of the fifteenth century, the Spanish rabbi Abraham Zacuto

11. Frisius was 250 years ahead of his time with this suggestion. The clocks of his day were far too inaccurate for the purpose, though the idea of using time differences to estimate longitude were sound, and would eventually be adopted.

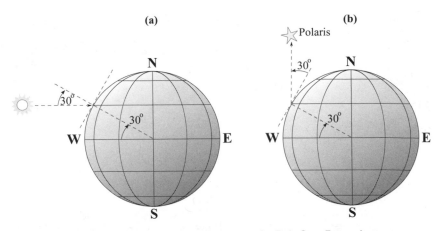

FIGURE 6.11. Determining latitude from the sun or the Pole Star. From elementary geometry, an observer's latitude is measured by "shooting" the sun or the Pole Star. (a) At equinox, if the sun is due south at noon, a measured altitude of 60° (with the sun 30° from zenith) means that the latitude of the observer is 30°. (b) The measured altitude of the Pole Star equals the observer's latitude, apart from a correction for the deviation of Polaris from true north.

improved the mariner's astrolabe, and it became quite accurate.[12] The improved mariner's astrolabe was used to "shoot the sun"—to measure its angular altitude for the purpose of estimating latitude. The instrument was suspended at waist height, and its alidade adjusted so that sunlight passed through a small aperture at the upper end and lit up the lower end, thus ensuring that the alidade was pointing to the sun. The solar elevation angle or altitude was read off the scale, with an error of less than a degree— unless the navigator was on board a pitching or rolling ship. The mariner needed to compensate for certain errors of measurement, and that entailed consulting tables or performing mathematical calculations. For example, if the measurements were made at night and the altitude of Polaris was measured, compensation had to be made for the star's small movements (these details of celestial wanderings were appreciated at the time of Columbus). But such tables and calculations were not always required. While the sun is at its noon zenith during an equinox, its altitude is simply related to the ship's latitude, as you can see from figure 6.11. In addition, a navigator fortunate enough to have a clear sky and a calm sea, so that he could

12. Rabbi Zacuto left Spain following the expulsion of Jews in 1492 and found sanctuary in the Portuguese court of John II, successor to Henry the Navigator. Here, he developed the improved astrolabe.

shoot the noonday sun, also obtained his heading without calculations because, of course, the noon sun is directly south for an observer in the northern hemisphere.

Much closer to home, a pilot often needed to estimate his distance from a nearby shore. If he knew that rocks, a reef, or a sandbar lurked just beneath the surface, then learning how far he stood offshore became a matter of some importance and urgency to an anxious mariner. As sea traffic increased, especially off the shores of newly found lands with coast-lines that had not been completely surveyed, such as the Spanish Main (the Caribbean Sea and the Gulf of Mexico), this problem was exacerbated. There are a couple of very old rough-and-ready techniques that have been employed for centuries by men of the sea to gauge how far a ship stands off from shore. Both methods require the mariner to extend his thumb at arm's length. The width of a thumb at the end of an arm subtends an angle of about $1.6°$. Suppose a pilot sees a church steeple, a tree, or some other object of known height on the shore. He can hold out his thumb horizon-tally and estimate the size of the steeple or tree compared with his thumb.

To provide a specific example germane to this period, let us say that a Spanish galleon is returning home from Cartagena and is passing along the northern coast of South America, perhaps laden with silver, in the year 1550. From his charts, the pilot knows that a coastal cliff he is passing is 50 m high. He also knows that just beyond this promontory, a hidden reef extends 2 km into the sea. He judges that the cliff height is half the width of his extended thumb, held horizontally. Does he need to steer a course further out to sea? No, in this case, assuming that he does not have a particularly fat thumb: from the trigonometry we can see that the cliff is more than $3\frac{1}{2}$ km distant.

The second thumb method is equally venerable: it is still used today by weekend navigators as a handy (pardon the pun!) technique. The *eye-blink* method makes use of the idea of *parallax*—that separated observers looking at the same object see a different background. In this case, the pilot knows that two coastal features—say, a lighthouse and a small islet—are 1 km apart, and he wants to estimate his distance from the coast. This time he extends his arm with the thumb up. He closes his left eye and, with his right eye, lines up his thumb with the lighthouse. He then closes his right eye and sees how far toward the islet his thumb appears with his left eye. Let us say that his thumb as viewed through his left eye is exactly lined up with the islet. An arm length is about ten times the distance between a pair of eyes; then, from elementary trigonometry we know that the angle between two lines extending from the ship to the lighthouse, and from the ship to the islet, is about 6°. Our pilot knows his geometry and estimates that he is about 10 km from the coast.[13]

Dead reckoning, at least in its early days, was every bit as approximate as these simple thumb methods.

Dead Reckoning

Navigators had caged latitude since classical antiquity, but as we have seen, longitude remained wild and free, and dangerous because of that. Longitude would not be tamed until the eighteenth century. At the time that Spanish and Portuguese mariners first set out into the wider world, longitude was estimated by dead reckoning (DR).[14] Whenever a caravel or a galleon found itself in open water out of sight of familiar landmarks, the navigator made use of dead reckoning to estimate the position of his ship or, more precisely, how far and in which directions it had traveled.

We have seen how Henry the Navigator established at Sagres an institution that nowadays would be called a navigation research center. This was necessary because mariners knew perfectly well that their skills, honed along the coastlines of Europe and particularly in the Mediterranean Sea, would not be good enough for the open ocean. In the Mediterranean there were no tidal streams, currents were known, a ship was seldom out of sight of land for more than a few days, and lines of longitude were separated by about the same distance because the Mediterranean extends east-west much more than north-south. Portolan charts for the Mediterranean region were well developed by the fifteenth century. All of these facts made for relatively easy navigating. Out in the open ocean, however, it would be a different story.

Dead reckoning estimated the path of a ship by determining the distance and direction traveled each hour and then adding up all the distances and directions. The idea is shown in figure 6.12a. The same method, called

13. Modern mariners can find equivalent tricks—cheap-and-cheerful methods—for roughly estimating the distance to shore in many of the popular sailing books. For example, if you have normal vision and can just discern individual trees on a shoreline, they are about one nautical mile away. If you are sitting on deck and can just see the line where land meets sea, you are about three nautical miles away. These two distance-estimation methods are based on average eye resolution and the earth's curvature, respectively.

14. The odd term, according to the *Oxford English Dictionary*, arises from an old-fashioned adjectival use of the word *dead*, meaning "complete, unbroken, or unrelieved," as in "dead calm." Thus, "dead reckoning" refers to a method that involves continuous calculation.

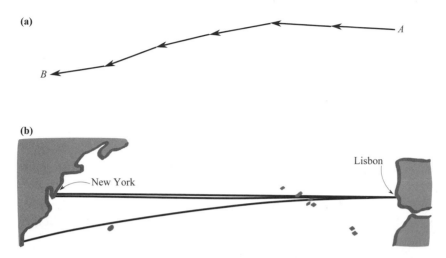

FIGURE 6.12. Dead reckoning. (a) A ship's route is estimated by making speed and direction measurements, and so producing a vector (an arrow) representing one hour's movement. Adding up the vectors for the movement of many hours produces the ship's route. Open ocean mariners performed such dead reckoning measurements and calculations every hour 24/7. (b) Three hypothetical voyages navigated by dead reckoning from Lisbon to what is now New York, in the sixteenth century. The lowest path is subject to a systematic error of ¼° in bearing estimate, unlike the other two paths, which are subject only to unbiased bearing estimate error of ¼°. All three ships are trying to head due west, which means that the total distance is 5,535 km. (Had they known to travel a great circle route, somewhat further north, the distance would have been shorter by about 100 km.) The dead reckoning measurements here are simulated on the assumption that measurements are made daily and that speed is 100 nautical miles per day (7.7 km/h).

path integration by scientists, is still used today. The implementation is now automated and is much more sophisticated, as we will see, but the idea is exactly the same: add up path segments. In the late fifteenth century, when Columbus and colleagues were heading west (and south), they made use of dead reckoning and obtained the necessary measurements with a mariner's compass; an *ampoletta* (an hourglass); and a mariner's astrolabe or a quadrant. We have seen how angles were measured, but dead reckoning requires distance estimates. How were these obtained?

The speed of a ship was estimated by throwing a wood chip overboard and measuring with an hourglass[15] the time the chip took to pass between

15. In fact, I should probably call the instrument a *marine sandglass* because the version used to estimate ship's speed would not last for an hour before being turned. For estimating ship's speed, it would empty in about 30 seconds (28 s became standard when chips logs were developed).

two points marked on the ship's rail. One would then multiply the speed obtained by one hour to obtain the distance traveled in that hour. Direction was measured with a compass, and, to use modern terminology, a trajectory vector for the ship was obtained. These vectors were joined together, tail to tip, to obtain the trajectory of the ship, as shown in figure 6.12a. In Columbus's day the concept of vectors did not exist; mariners stored the measured data on a *traverse board*, a circular board with the points of the compass marked. Each hour a peg was placed on the board on a radial line corresponding to the measured heading and at a distance from the center that represented the calculated distance traveled in the last hour. After 24 hours, the navigator would add up all the distances and directions (calculate the integrated path) to estimate the day's travel and would then remove the pegs to make the board ready for the next day.

As you can imagine, dead reckoning was, especially in its early days, a very rough-and-ready method for estimating location. The compass measurement of direction was subject to error, and the estimate of time and therefore of distance was very approximate. Worse, there may have been systematic errors due, for example, to a current or tidal stream. Or perhaps wind and current were sensed but compensated for incorrectly. Some errors would be random and unbiased, which is to say that the estimate of speed would be too large just as often as it would be too small, so that the average estimate would be about right. More insidious were systematic errors, which were biased in one direction and accumulated over time.

To illustrate the effect of dead reckoning errors, consider figure 6.12b. Consider yourself a not-too-skillful chief navigator, heading out from Lisbon to the New World—say, at the beginning of the sixteenth century. You have three ships under your command and are planning to sail due west in order to land on North American soil at a place fairly close to what would become New York. Each ship measures speed and direction daily, not hourly, with a maximum random error of, let us say, 5% for distance and a quarter degree for direction—pretty good for the period. Let us say it is foggy the whole way; your three ships become separated and lose contact with each other. I can simulate the paths taken by your ships on a computer using dead reckoning, with the errors just quoted.

Two of your ships follow the more northerly courses shown in figure 6.12b. They pass through the Azores and end up reasonably close to modern New York—within a few tens of miles. The error in estimating distance is not too serious in this instance; the sailors know they have reached the New World when they bump into it. The unbiased direction errors result in small deviations from the desired direction, as we can see from the figure. You are navigator for the third ship. There is a fault with your mariner's compass, or with the way you read it, so that, unknown to you, your measurements of bearing have a systematic error of a quarter degree. That is, each day your heading is further south than you think it is by a quarter degree. You are trying to steer a course due west but end up passing Bermuda and reaching North America at Florida. Such are the dangers of accumulated, systematic errors.

Of course, the real world is not as simple as my simulation suggests. Currents will vary; there will be differences in the magnitude of both unbiased and systematic measurement errors; storms will disorient you and blow you off course. On the other hand, you will also be able to compare notes with other ships in your fleet for some of the time and will be able to make better measurements or reset your dead reckoning path integration when you make landfall, as at the Azores in my illustration.

The chip log was first introduced in the last quarter of the sixteenth century. It replaced the older chip of wood and made the estimation of ship speed more systematic. A standard shape evolved over the years: the log was shaped like a slice of pie, as you can see in figure 6.13, and was weighted on the curved surface so that this surface would be oriented to the bottom when thrown in the water. The log was attached to a long knotted rope which spooled out from a reel as the log receded from the moving ship. The knots were regularly spaced; the number that were counted out in a given time interval (measured by a sandglass) provided an estimate of speed— hence, our term *knot* for speed measured in nautical miles per hour. While providing better estimates of a ship's speed through the water than the old wood-chip method, the chip log still suffered systematic errors from, for example, sea currents. Speed over water was well measured, but it is speed relative to the land that matters when estimating ship movement.

Sir Francis Drake

Born a commoner, the eldest of 12 children of a laborer from a small provincial town in southern England, Francis Drake was an unpopular parvenu with many of his contemporaries, such as Martin Frobisher. Drake earned his living as a privateer—a polite word for a state-licensed pirate— for 10 years, plundering the Spanish Main on behalf of his queen, Elizabeth I. Returning to England from the Caribbean in 1577 (the same year that the chip log was introduced), he was sent to the South Sea—the

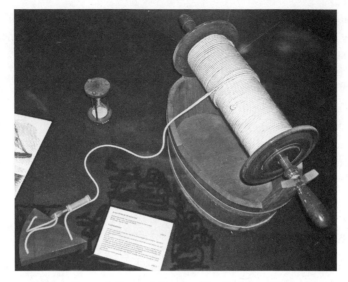

FIGURE 6.13. Chip log (*lower left*), sandglass, and knotted line: the equipment required for dead reckoning of a sailing ship route. Dead reckoning in this manner lasted into the nineteenth century. Photo by Rémi Kaupp.

Pacific—to plunder Spanish possessions once again, in what would become an epic three-year journey.

Drake set out with five small ships and between 164 and 200 sailors plus a "mob of adventurers."[16] His flagship, the *Golden Hind*, which was an English version of the galleon (fig. 6.14), had a crew of fewer than 100 men. The small flotilla made its way down the eastern coast of South America. In Patagonia, Drake executed a plotter (interestingly, in the same region where Magellan had had to deal with a mutiny half a century earlier). Two of the ships were abandoned near the River Plate; the remaining three passed through the Strait of Magellan, but two of them missed a rendezvous with *Golden Hind* on the western side. These two passed back through the strait and made their way to England with the news that Drake was lost.

Far from it: Drake was fulfilling the purpose of his expedition by plundering Spanish coastal towns and harbors on the coasts of Chile and Peru, easily taking Spanish ships and their valuable cargoes. Because the Span-

16. Owen (1979). Much has been written about Francis Drake and his famous expedition. See, for example, Bawlf (2003); *Encyclopedia Britannica*, s.v. "Francis Drake"; Owen (1979); and Thrower (1984). Herman (2005) includes a readable account of Drake's unsavory earlier life.

FIGURE 6.14. The *Vasa*, a Swedish galleon that sank in the Baltic Sea in 1628 and was recovered in 1959. This is the only original galleon still in existence. There were many galleon designs. The Spanish adapted one specifically for their treasure fleets. English galleons, such as Francis Drake's *Golden Hind*, were smaller and faster. Photo by Peter Isotalo.

iards were used to sole possession of these waters, they made few defensive preparations. Soon the *Golden Hind* was overloaded with gold, silver, and jewels. "El Draque," as he was called by his Spanish enemies, sailed further up the coast, leaving a monument at "New Albion," probably near San Francisco. He described the coastline here and made latitude measurements from his ship with a mariner's astrolabe, and perhaps while on shore with a larger and more accurate astrolabe. Much effort has been made in recent years to determine the exact whereabouts of New Albion: basically, the uncertainty is due to the inaccuracy of Elizabethan navigational instruments, and in particular of the mariner's astrolabe. (Drake's estimates of latitude were accurate to perhaps 11 minutes of arc, which corresponds to a positional error of 20 km.)

Drake may have followed the coast as far north as Vancouver Island and, according to one controversial account, perhaps even farther north in an attempt to discover a northwest passage. He then sailed west across the Pacific, arriving in the Philippines after 68 days, in October 1579. The *Golden Hind* moved on to the Moluccas, where Drake made a treaty with the Moluccan sultan for trade in spices. By March 1580 he was in Java; he rounded the Cape in June and arrived back at his home port, Plymouth, on September 26, 1580, with 56 of his original crew. Drake was the first person to sail his ship around the world (del Cano changed ships after Magellan's death) and perhaps the first to circumnavigate the world from west to east. He was later knighted on board his ship, played a prominent role fighting the Spanish Armada, made a fortune from the slave trade, and died at age 50 in Panama (again on an expedition to fight Spaniards) of dysentery.

Of Drake's abilities, we have the word of his pilot. On his most famous expedition, Drake was fortunate in capturing a Portuguese captain, Nuno da Silva, off the Cape Verde Islands early in the voyage, in 1578. Da Silva knew the South American coast very well and was a great asset to the expedition. He described his new captain thus: "The first thing he did when he captured a vessel was to seize the charts, astrolabes and mariner's compasses. . . . He carries three books of navigation, one in French, one in English, and another, the account of Magellan's voyage, in a language I do not know. . . . He is a very skillful mariner."[17] Da Silva was freed after 15 months' captivity, on the coast of Mexico, and was later questioned by the Inquisition about El Draque.

17. Thrower (1984), p. 182.

Great Circles

Another product of the school of navigation in Portugal during this period was Pedro Nunes, a great mathematician who wrote the first mathematical treatise on navigation—in the same decade that Frisius was contemplating longitude. Frisius's student Mercator realized that the sailors of his day were wrong in assuming that a straight line (shortest-distance) path across the oceans required following a fixed compass course. Nunes had described the *shape* of the fixed-compass course, the so-called rhumb line, or loxodrome:[18] it spiraled toward the poles because of the converging meridians. In 1594 John Davis (the same Davis who has a quadrant named after him) showed how mariners could sail a great circle route—the true shortest path between two points on the surface of a sphere—by constantly adjusting their bearings. This technique is obviously much more difficult to implement in practice than simply following loxodromes, which were straight lines on a Mercator map.[19]

Some explorers may have known how to follow a great circle route before Davis formally outlined the method. For example, Sebastian Cabot seems to have followed such a route a century earlier in his explorations in the North Atlantic.[20] (At the high latitudes of his expedition, great circle routes can differ markedly from rhumb lines.) Davis's achievement was to publicize the method so that everyone could take advantage of it. By the last decade of the sixteenth century, the mathematics of Mercator projections, rhumb lines, and great circles had been worked out, so that all navigators knew how to change bearing depending upon their position and thus follow a great circle, the shortest route between *A* and *B*.

Navigation was changing, moving from an art to a science. Navigational instruments would follow a parallel course, from being the products of a craft tradition to being the products of precision engineering. All the time, empirical knowledge was being systematically built up. Now the French got in on the act by developing *routiers*, known to the English as *rutters*—a

18. Technically, *rhumb line* refers to the course and *loxodrome* to the mathematical curve.

19. Mercator knew of Nunes's work, which appeared in a 1537 treatise, and studied it in detail at Louvain. Edwin Wright, who was from the next generation of navigation theorists and provided the mathematical underpinning for the Mercator projection, was also much influenced by Nunes.

20. Sebastian Cabot was the son of John Cabot, the official European discoverer of North America, in 1497. John Cabot also returned to his home port of Bristol, in southwest England, with confirmation of the existence of rich fishing grounds—the Grand Banks off Newfoundland.

Navigation Meets Science

Crusading knights returned from the Middle East with new knowledge acquired from Islamic peoples, and from that time many of the famous universities of Europe can be dated: Padua, Paris, Oxford, and Verona. By the fifteenth century, educated people were beginning to lose faith in the wisdom of the ancients and in conventional beliefs. After all, medical wisdom had been shown to be woefully inadequate during the recent Black Death pandemic; Ptolemy was wrong about the impossibility of reaching India by sea; the recent introduction of gunpowder showed that substances could be combined in ways unforeseen by the ancient philosophers. Gunpowder, indeed, was a product of practical experimentation, not philosophy—so perhaps such experimentation could lead to advances in other areas, such as navigation. New ideas could be promulgated faster than ever before through another new invention: the printing press. Europe was gearing itself up for the scientific revolution of the seventeenth and eighteenth centuries.

navigational database (to use a modern term) very similar to the Italian portolan charts. Such data and charts were combined beautifully, in what many pilots at the time and for generations afterward considered the definitive form, by a Dutch cartographer, Lucas Waghenaer. His book, translated into English in 1588, was referred to simply as *Wagonners*, and it guided many an English navigator across the Atlantic to the newly formed American colonies.[21]

Mariners were learning to sink the land, and yet, despite the disorientating uniformity of the open ocean, they could make a reasonable estimate of their latitude and bearing, though not yet their longitude.

21. For more on this interesting period of navigation and exploration history, see Arnold (2002), Bawlf (2003), Cunliffe (2002), Dunn (1989), Hakes (2009), Kelsey (2000), Love (2006), Mason (1962), Papenfuse and Coale (2005), and Thrower (1984). I especially recommend Bedini (1998). For a detailed account of the scramble for control of the Spice Islands, see Milton (1999).

CHAPTER SEVEN

The Age of Sail and Steam

The seventeenth through nineteenth centuries saw the emergence of new colonial empires—the Dutch, the French, and especially the British—and new maritime navigational techniques to assist the flow of trade goods from colonies to home countries. In this chapter we see how the sextant was employed by maritime navigators and how the longitude problem was solved. Land-based exploration and navigation also increased, as continents were explored and mapped for the first time.

Bearing Up

Vernier introduced his scale; magnetic variation was discovered and mapped;[1] galleons gave way to barks, brigs, and sloops. All of these developments were important, but most important for us during this period was the introduction of the sextant for surveying and navigating. This instrument permitted estimation of azimuth (horizontal) angle as well as elevation angle, unlike earlier devices such as the mariner's astrolabe, and could therefore be used to provide a bearing. Mariners used increasingly refined sextants for centuries, not only to estimate their positions at sea but also to survey the world's coastlines. Sextants are still part of a navigator's toolkit, if only as a backup for satellite navigation.

Here I provide some examples of sextant use in coastal navigation that show the increasing sophistication and accuracy of position-estimating techniques in general, and in particular of the techniques for estimating

1. Edmund Halley mapped magnetic declination, and Henry Gellibrand published yearly accounts of the changes, both in the seventeenth century. See Wikipedia's entry "Earth's Magnetic Field Declination" for an interesting animation of the change in declination over the last four centuries.

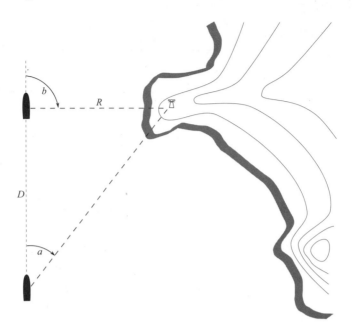

FIGURE 7.1. Estimating distance off. By measuring bearing angles *a* and *b* to a prominent reference point onshore—here, a coastal beacon—and then estimating the distance, *D*, covered by his ship between the two measurements, a navigator can estimate *R*, his distance off.

"distance off"—the distance of a ship from the shore. There is a stark contrast between these processes and the rough-and-ready methods of earlier times.

Consider the ship of figure 7.1. (Let us imagine it to be an eighteenth-century French bark, exploring the Pacific Islands.)[2] With his chip log and pocket watch, the navigator can measure the distance, *D*, that his vessel covers between two angular measurements. These angles, *a* and then *b*, obtained with a sextant, are between the ship's heading direction and a prominent coastal feature, in this case a beacon. From the geometry our navigator can calculate *R*, the distance off. For example, if he determines that angle *a* is 40° and *b* is 90°, and that the distance *D* between the two measurements is 2.00 km, then he finds that *R* is 1.68 km.

2. During this period the French "discovered" many of the 20,000–30,000 Pacific Islands, as free-roaming captains such as de Bougainville planted the French flag on any remote speck of land in this vast ocean. Their great rivals of the age, the British, did likewise, of course. Earlier, the Spanish had been at the fore of Pacific exploration, with three expeditions from South America that reached as far as the Solomon Islands, off the east coast of New Guinea, and the New Hebrides.

Our intrepid navigator might equally well employ a sextant to determine his position relative to the coast by referring to two coastal landmarks, in which case he does not need to estimate his ship's speed because this method works for a stationary vessel. In figure 7.2 we see that he is at anchor and measures angles *a* and *b* from a common reference direction (magnetic north, say) to two landmarks, the beacon and a large palm tree atop a small hill. On his local chart he marks a *line of position* (LoP) for each direction from each landmark. The intersection point of these two LoPs indicates the ship's position.

In general, if a navigator has been estimating his position by dead reckoning (perhaps he has been in open ocean for some time, with no terrestrial reference points), then he might confirm his estimated position upon sighting land by making a bearing measurement: the intersection of one LoP with his dead reckoning track gives his position. Of course, this method assumes that the landmark he spots was at a known location, marked on the navigator's charts. For training purposes, a navigator might ask his midshipmen to estimate the ship's position by dead reckoning as it weaves its way through an atoll of Pacific islands and then

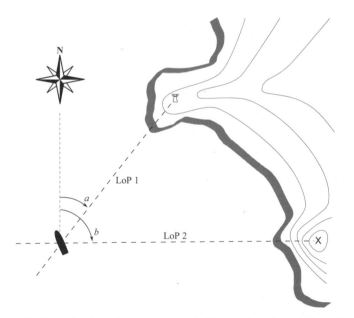

FIGURE 7.2. Getting a fix. A navigator or coastal pilot obtains a fix on his position relative to a charted coast by measuring two lines of position—the compass direction to two known landmarks or reference points on his chart. The intersection of the two LoPs shows the navigator his position on the chart.

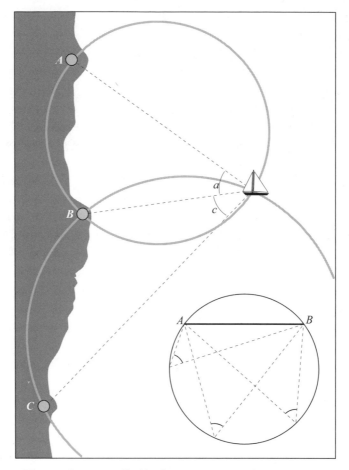

FIGURE 7.3. Three-point sextant fix. You have an accurate chart, a sextant, and a knowledge of geometry, and use these to fix your yacht's position. By measuring the angle *a* between reference points *A* and *B*, you know that your position is somewhere on the circle that passes through both these points. Why? The inset shows a geometrical theorem: all the marked angles are the same and so the ship must lie somewhere on the circle. (Interested readers can calculate the circle radius in terms of the angle and the distance *AB*.) By measuring the angle *c* between *B* and *C*, you can draw another circle; the yacht's location is the seaward intersection of the two circles.

check the accuracy of their dead reckoning track by measuring LoPs from different islands.

Fast-forward to the very near future. You have just won the lottery and are now retired, living aboard your luxury yacht off the coast of Australia. You want to sail the Great Barrier Reef, but you are aware that these are dangerous waters: even Captain Cook nearly lost his ship on the coral. To confirm your GPS position, you bring out a sextant and do it the old-fashioned way (a "three-point sextant fix" method that dates from 1775), making use of those marvelously detailed and accurate charts that earlier generations of seafarers had risked their lives to create after making arduous journeys and many thousands of survey measurements. Your charts indicate the distances between three landmarks, A, B, and C, that you see along the nearby coastline (fig. 7.3). By measuring two angles between these landmarks, and using some navigator's geometry, you fix your position. The method is summarized in figure 7.3.[3]

Captain Cook

James Cook lived in a time of scientific enlightenment (which kick-started the Industrial Revolution). He had been appointed by the British Admiralty to lead a scientific expedition to Tahiti to observe a rare astronomical event, the transit of Venus, which was predicted to occur in June 1769.[4] A *transit* occurs when a planet passes in front of the sun. In this case the scientific community wanted three observations from widely spaced locations on Earth (Tahiti, the North Cape of Norway, and Hudson's Bay). In each location, scientists would measure the time taken for Venus to cross the sun, and from these measurements it was hoped that the scale of the solar system could be estimated—in particular, the distance from the earth to the sun. (It turned out that the measurements from all three locations were insufficiently accurate to provide a meaningful estimate of this distance.)

Cook and his ship, *Endeavour*, were both unusual for such an expedition. The Admiralty was happy to carry out this scientific mission because it had an ulterior motive, as we will see. *Endeavour* was not a warship, but a humble collier, converted for Cook's first expedition (fig. 7.4).[5] However, she was very strong and an excellent sailor. The 40-year-old Cook was from equally humble origins, and his maritime credentials had been established

3. The three-point sextant fix method is discussed in, e.g., Mills (1980). Australia had been discovered by the Dutch and British; later, Cook was prominent among the explorers who provided detailed surveys of the coastline. Indeed, Cook had earlier made his name in North America on the basis of his surveying skills.

4. Much has been written by and about James Cook. See, for example, Allen (1980), Brown (2003), Herman (2005, chap. 13), Holmes (2010), Owen (1979), and Price (1971). For more on the transit of Venus measurement, see Danson (2006).

5. *Endeavour* was a *bark*—a sailing ship with three masts, square-rigged on the front two and fore-and-aft rigged on the aftermost (mizzen) mast.

FIGURE 7.4. HMS *Endeavour*, painted in 1790 by Thomas Luny. The ship is depicted as she was in 1768, before Cook renamed her. She was extensively refitted and strengthened for his first voyage. Wikipedia.

while he was in the merchant marine, not the Royal Navy. Despite these perceived shortcomings, he was appointed leader of the expedition because of his well-known ability as a surveyor and navigator. Throughout his voyages he had surveyed coastlines very carefully. He took "sun-shots" with a sextant to establish latitude and made lunar observations, accompanied by many mathematical calculations, to establish longitude.[6]

Cook set out in August 1768 after much preparation, with many provisions,[7] and with scientists, most notably Joseph Banks, a wealthy botanist who would make his name with Cook. The scientists would bring back descriptions, drawings, and samples of many plants previously unknown to Europeans, as well as of animals, insects, birds, and fishes. *Endeavour* headed south, passed Cape Horn without incident, and arrived in Tahiti. At this point the crew had suffered not at all from scurvy (the scourge of mariners from the Age of Exploration, when long sea voyages were first

6. On his second expedition Cook benefited from a Kendall marine chronometer—a copy of John Harrison's award-winning design, following his epic struggle with the Board of Longitude to design a robust and accurate timepiece that would work on a moving ship. This chronometer greatly simplified the estimation of longitude. As we will see, Cook tested the chronometer and found it to work very well. See Sobel (1996).

7. *Endeavour* was loaded with 94 people; 17 tons of biscuit; 5 tons of flour; 2,500 lb of raisins; 1,500 lb of sugar; 500 gal of vinegar; 1,200 gal of beer; 1,000 gal of brandy; 1,000 lb of desiccated soup; 4 tons of sauerkraut; and salt meat for 12 months.

undertaken) because of Cook's enlightened ideas. He experimented with different foods, such as sauerkraut and fresh fruit, and insisted that each crew member partake his share.

After the transit was observed, Cook opened secret Admiralty orders which instructed him to head further south and search for the supposed *Terra Australis Incognita*—the unknown southern continent that Ptolemy, sixteen centuries earlier, had supposed existed. If he failed to find this new continent, Cook was to explore Australia.

He circumnavigated and charted New Zealand for five months and then, no new continent having been found, carefully surveyed 2,000 miles of eastern Australia (during this time Banks named Botany Bay). *Endeavour* struck the very treacherous Great Barrier Reef, eventually floating free. Cook wrote in his journal: "To those only who have waited in a state of such suspense, death has approached in all his terrors. . . . Every one saw his own sensations pictured in the countenances of his companions; however, the capstan and windlass were manned with as many hands as could be spared from the pumps; and the ship . . . was heaved into deep water."[8] On the way back to England, *Endeavour* put in to the Dutch colony at Batavia (Jakarta) for much-needed and extensive repairs, during which time 40 of the crew died from malaria or dysentery (but none from scurvy). Cook arrived in Plymouth in June 1771.

Cook conducted a second voyage of exploration from July 1772 to July 1775, again searching for the famed ice-free southern continent, this time establishing that it did not exist. He charted the lands he did encounter, including the island of South Georgia (fig. 7.5a). He discovered Easter Island (for Europe) and returned home after traveling 70,000 miles. The third expedition was intended to find the Northwest Passage across the top of Canada. For these explorations he wintered in Hawaii, where he was killed by indigenous people. He is remembered for providing a comprehensive map of the Pacific, for banishing the persistent myth about Terra Australis Incognita, for leaving detailed journals, and for bringing home much information about the people, the animals, and the plants on the other side of the world. As with Columbus, Cook's likeness is preserved in statues all over the world (see fig. 7.5b for an example).[9]

8. Brown (2003).

9. According to the Captain Cook Society website, there are at least 23 Cook statues around the world—in Australia, Canada, New Zealand, Russia, the United Kingdom, and the United States—plus hundreds of other memorials such as library and building names, street names, plaques, and columns.

(a)

(b)

FIGURE 7.5. A Cook impression and an impression of Cook. (a) The island of South Georgia, located about 1,900 km east of the Falkland Islands and a similar distance from Antarctica. The island was charted by Cook during his second voyage, which ended in 1775; this map was published in 1777. Note that Cook's track is marked and that, on this map, south is up. (b) A statue of Captain Cook on the Hawaiian island of Kauai. (a) From Cook's *Chart of the Discoveries made in the South Atlantic Ocean, in His Majestys Ship Resolution, under the Command of Captain Cook, in January 1775*, published by W. Strahan and T. Cadel. (b) Photo by the author.

The Cocked Hat

The tricorn hat, popular in the eighteenth century, gave its name to the error triangle that results from incompatible angle readings, when converted into LoPs and drawn on a chart. Let us say that a certain captain Jean-Luc Picard (a distant relative of the famous surveyor of France) of the French brig *Enterprise* has made three readings with his sextant and is now staring at the three corresponding LoPs that he has drawn on his chart of French Polynesia. Two lines intersect, indicating his ship's location, but the good captain has decided to double-check and has made a third reading. Unfortunately, the LoP for this third reading does not intersect at the same point as the other two LoPs, as shown in figure 7.6a. Because of measurement error, none of the positions of the LoPs is exactly right, and the three LoPs show three intersection points that define a triangle (the cocked hat).

Which of the intersection points is the right one? Probably none of them. It was generally assumed in the eighteenth century that the true position of the ship was somewhere inside the cocked hat, but this is not necessarily correct, as we will see.[10] It can be shown from elementary statistical arguments, for example, that the probability of the ship being inside the cocked hat is 0.25, which means that there is a 75% chance that the ship is outside. The analysis leading to this conclusion assumes that the error is purely random and unbiased (i.e., that there is no systematic error, and there is an equal chance of the ship's true position being on either side of any LoP). The probabilities of finding the ship in any of the other areas outside the cocked hat are shown in figure 7.6a. While this argument shows that the most likely place for the ship is indeed within the cocked hat, this likelihood is small.[11]

Suppose that Captain Picard is aware that he is not very good at using a sextant and that, in particular, he is aware that he always reads the angle incorrectly—say, he consistently overestimates the angle by one minute of arc. This systematic error converts into shifted LoPs, as suggested in figure 7.6b. He can correct for this known error, and the result is a much

10. The phrase "knock into a cocked hat" derives from the corrections applied to mitigate this error of navigation, though its meaning has shifted since it was first coined.

11. Cocked hats are still a part of navigation today, and the technique is described in many practical handbooks, such as Bartlett (2009, p. 59), Karl (2004, pp. 131–32), and Toghill (2003, p. 65). The statistics of cocked hat positioning errors have been extensively analyzed; you can read about them in Anderson (1997), Daniels and Wishart (1951), and Hiraiwa (1967).

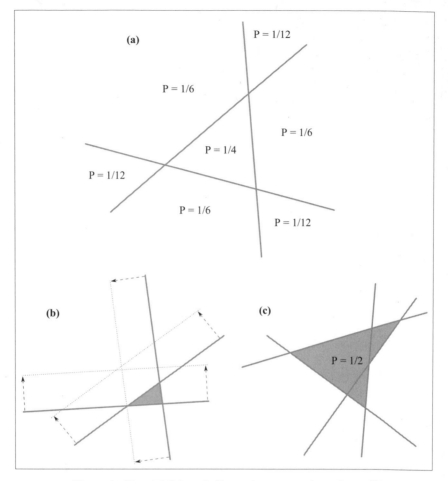

FIGURE 7.6. The cocked hat. (a) If three LoPs are drawn on a chart, they will intersect only at a single point for the case of zero measurement errors. In general, they intersect as shown; the central triangle is called a cocked hat. The probability for the ship's being inside each area is shown, assuming unbiased, random measurement errors. (b) A known systematic error can be corrected as shown. The new cocked hat (defined by dotted lines) is much smaller, and in this case is outside the original cocked hat (shaded). (c) For four LoPs, the probability of the ship's being inside the shaded area is 50%.

smaller cocked hat, so that our captain has reduced the size of the most likely location of the ship—he has reduced his position estimation error.

If a fourth measurement is made and a fourth LoP added to the chart, then the same type of statistical analysis as before (assuming zero systematic error and no bias) leads to the conclusion that the probability of the ship being inside the quadrilateral of figure 7.6c is 0.50. In general, for more and more LoPs the probability of the ship being inside the largest bounded area increases. In fact, modern statistical techniques replace the bounded area (be it a cocked hat, the quadrilateral of fig. 7.6c, or whatever) with a *confidence ellipse*, which is drawn in the region of the intersection points (by a computer program, nowadays). The probability of the ship being inside the confidence ellipse is some stated value, such as 95%. Several such confidence ellipses can be drawn, like contours: the 99% ellipse will be larger than the 95% ellipse and will be outside it; the 50% ellipse will be smaller than the 95% ellipse and will be inside it.

The Longitude Problem

In the early 1700s the single greatest obstacle to accurate navigation at sea was the same as in Columbus's day, 200 years earlier. We will soon see that for navigators on land during this interval, significant progress had been made using celestial observation methods. But these methods could not be applied on the pitching and rolling decks of ships on the open ocean, so, when it came to determining longitude, maritime navigators were all at sea.

To solve the longitude problem, the various crowns of Western Europe —those with extensive overseas possessions or trade links—offered large prizes. The first to do so was Phillip II of Spain. (By no coincidence, Spain was the first nation in Europe with an extensive overseas empire.) In 1567 Phillip offered a prize of 7,000 ducats plus an annual pension of 2,000 ducats to the person who could provide a method for accurately determining longitude.[12] In 1598 his successor, Phillip III, increased the value of this prize. The Dutch offered a large cash prize in 1636, at a time when they were establishing their preeminence in the spice trade. The British and French overseas trading links, and then empires, arose somewhat later, so for them the urgency of estimating longitude at sea became overwhelming only in the early eighteenth century. The British established their Board of Longitude in 1714, offering up to £20,000 to the person who invented a method for accurate determination of longitude at sea.[13] The French

12. A ducat was a gold coin used for international trade that weighed one-eighth of an ounce. In today's money, one ducat was worth about $140.

13. The British currency of 1714, when the Longitude Prize was first offered, was worth the equivalent (based on the price of gold) of $320 today, so that £20,000 would be worth over $6 million.

FIGURE 7.7. The Scilly Isles naval disaster of 1707, as depicted by an unknown eighteenth-century artist. Wikipedia.

Académie Royale des Sciences offered a valuable *Prix Rouillés* for navigation a year later.

The inability to estimate longitude at sea was not merely an inconvenience; it was dangerous. Many a ship had been lost at sea through the crew's ignorance of their true position. Most infamous of these losses, and the immediate spur to the creation of the Board of Longitude and the Longitude Prize in Britain, was the disaster of 1707, one of the greatest in British naval history. A large fleet of 21 warships, under the command of the ludicrously named Sir Cloudesley Shovell, was returning home to England from Gibraltar. In bad weather and with very poor visibility, the fleet ran aground on the rocks of the Scilly Isles, a group of islands to the west of Cornwall in the extreme southwest of England. Four ships foundered, including three ships of the line (fig. 7.7). The total number of dead was estimated at between 1,400 and 2,000, including the commander-in-chief, whose body was washed up on shore and recovered a few days afterward. A government inquiry concluded that the reason for the disaster was the pilots' lack of knowledge of their true position owing to their inability to accurately estimate longitude.

The Board of Longitude was set up as a direct consequence of this disaster, to oversee and judge potential solutions to the problem that were proposed and brought before them. Anyone who had a sensible suggestion (and more than a few with nonsensical suggestions, as it turned out) could be considered. The prize was specified in detail: if a reliable solution provided a ship's navigator with an estimate of longitude that was in error by no more than one degree of arc, then the prize would be £10,000. For a solution accurate to within 40' of arc, the prize would be £15,000, and for a solution accurate to within 30' of arc, £20,000. (One minute of arc corresponds to a distance of one nautical mile.) In addition, the board could award discretionary benefits and expenses incurred, including advances for anticipated expenses.

Such was the extent of the problem—or, to be more precise, the difficulty of providing a workable solution—that the board sat for more than a century. It finally was dissolved in 1828, and by then there were two methods that were known to work, and had been known to work for half a century. The reason for the long interval between finding a reliable method and disbanding the board was that the solutions needed to be tweaked in order to make them affordable enough that every ship venturing into open ocean would be able to determine its latitude and longitude.[14]

Measuring longitude accurately is equivalent to measuring time accurately because the earth rotates at an angular rate of 15° per hour.[15] Thus, to take a simplified hypothetical example, let us say that a navigator on board an English ship in the Mediterranean Sea looks at his pocket watch (which is perfectly accurate and which he has not adjusted since he left home port) and sees that local sunrise occurs at exactly 7:20 a.m. Greenwich Mean Time. He consults an almanac and reads that, for this day of the year, sunrise at Greenwich, London, occurs at exactly 6:00 a.m. GMT. Thus, the ship is 80 minutes ahead of Greenwich. From the rate at which earth rotates, we see that this time difference converts to a longitude difference of exactly 20°. The navigator determines that his ship is 20° east of Greenwich.

You see the catch: to estimate his longitude, the navigator needs to know the time at Greenwich more accurately than any pocket watch of the day could achieve. He can measure the Greenwich Mean Time in a number of different ways, and we will unpack these different methods over the next few sections. Note also that a reference longitude is required. It took a very

---

14. The story of the longitude problem and its solution has been told many times. The interested reader should consult Howse (1980), Sobel (1996), Taylor (1971), and especially Andrewes (1993). There are also many articles online.

15. I might also refer to the angular rate of the earth, confusingly, as 15 seconds per second, or $15'' \, s^{-1}$. That is, the earth rotates through 15 seconds of arc per second of time.

long time for the nations of the world to agree that the origin of longitude (the *prime meridian*) should be located at the Greenwich Observatory in London, England. Prior to that agreement, each country with maritime aspirations placed the prime meridian within one of its own cities (e.g., Cadiz, Naples, Paris). Such is the pride of nations that this agreement, ironically, took longer to hammer out than did the longitude problem itself: it was only in 1884, at a conference in Washington, DC, that international consensus was reached.

The main contenders for the longitude prizes were thus methods for accurately estimating the passage of time. The two successful candidates, to which we soon turn, were the marine chronometer method and the lunar distance method. In total, the British Board of Longitude awarded over £100,000 on various schemes during its century and more of existence. Of this total, some £23,000 was given, with some reluctance and considerable delay, to John Harrison for his marine chronometer; and another £9,500 went to other contributors to this method, including the watchmakers John Arnold, Thomas Mudge, and Thomas Earnshaw. The board awarded £5,000 to lunar distance contributors and £615 to Jesse Ramsden, whom we met earlier, for his dividing engine. Thus, the bulk of the prize was awarded to proponents of the marine chronometer, and so it seems as if this instrument provided the better solution. That would eventually prove to be the case, but in the late eighteenth and early nineteenth centuries, the situation was not nearly as clear-cut as it is sometimes made out to be today.

MEASURING LONGITUDE: EARLY CONTENDERS

In the seventeenth century two eminent astronomers had suggested solutions for the problem of longitude. Edmund Halley had proposed a method that considered the celestial background of the moon. He looked for stars that passed close to the moon or that were occluded entirely by it. He made observations in support of this method and tried it out at sea. It seems not to have worked, as he dropped the idea and nobody else advocated it. This method is similar in spirit to the lunar distance approach except that Halley restricted attention to stars that were, in angular terms, very close to the moon. Indeed, Halley became a supporter of the lunar distance method, and some of his observations contributed to it. Another idea of his involved the geomagnetic field. He hoped that magnetic deviation or variation (the difference between true north and magnetic north) might indicate longitude. In principle this might work if the magnetic field lines were

FIGURE 7.8. The Galilean moons of Jupiter. The four largest moons are, from the inside outward, Io, Europa, Ganymede, and Callisto. There are dozens of smaller moons that are visible only with a high-magnification telescope, not available in Galileo's day. Galileo realized that the relative position of these moons constituted a universal clock. He did not win a longitude prize with this method because it involved accurate measurement of the moons' positions, which was impractical at sea. Photo by Jan Sandberg.

fixed and were constrained to lie in a plane, but neither of these prerequisites applies in practice, so the idea was stillborn.

Somewhat earlier, in 1610, Galileo thought he might win the Spanish longitude prize with his idea of measuring time by observing the moons of Jupiter. He had discovered the four biggest Jovian moons (now called the "Galilean moons") with his newly developed telescope and realized that their regular orbits constituted a celestial clock that could be read by anyone with a telescope. It was not difficult to see these four moons even with the poor-quality telescopes of Galileo's time. (The moons are shown in fig. 7.8; they are readily observed through a modern telescope of modest magnification.) Giovanni Cassini, a century after Galileo, got the Jovian moon method to work well. By then, the orbits were known with sufficient accuracy to be able to predict them. Therefore, the positions of the moons relative to Jupiter and each other could indeed be used to tell the time and reveal the observer's longitude.

The method worked fine on land and was used well into the nineteenth century. An observer would take measurements and consult tables that showed the position of the moons at a given time in Greenwich, say. He would then check his local time, and from the difference he estimated his longitude. The trouble with this method was in making accurate measurements of the four moons while on the moving deck of a ship at sea. This problem proved intractable, and the method was therefore not adopted at sea.

FINDING LONGITUDE BY THE MARINE CHRONOMETER METHOD

The simplest method of measuring time is (no surprise) with a timepiece. Reliable pendulum clocks had been developed in the late 1600s, and these were accurate enough for the purposes of longitude estimation. But as was the case with the Jovian moon method, this approach was not practical on a ship. Pendulum clocks required a steady, unmoving base so that the pendulum oscillations were uniform. On a pitching and rolling ship—a sailing ship that would heel over at a considerable angle in crosswinds—pendulum clock accuracy went overboard. The great Dutch scientist Christiaan Huygens, working in France, tried for many years to get a pendulum clock to keep accurate time at sea, but never succeeded.

The first timepiece that retained its accuracy at sea required a different operating principle—a spring instead of a freely oscillating pendulum. This marine chronometer was the work of John Harrison, an English watchmaker of humble origins but lofty aspirations and stellar capabilities. Harrison completed a total of five handmade chronometers in his attempt to win the Longitude Prize; these timepieces are now conventionally labeled H1 to H5. The first three, which displayed increasingly sophisticated horological developments, all proved insufficiently accurate to merit submission to the Board of Longitude. H1 had been tested on a sea voyage from England to Lisbon and worked better than any earlier sea clock, but it did not satisfy the perfectionist Harrison. He obtained money from the board to develop H2 and H3, but these designs also did not work well enough. Harrison had begun work on H1 in 1730 and submitted H4 for consideration in 1761. More than three decades of intense effort produced a very small timepiece that was extremely accurate. In sea trials during a voyage from England to Jamaica, H4 was found to be only 5.1 seconds slow after more than nine weeks at sea.[16] The error in estimating longitude, based on this time error, was about one and a quarter minutes of arc (corresponding to a distance of about 2 km), which was well within the board's requirements for the maximum prize.

This prize was slow in coming; some members of the board favored the rival lunar distance method,[17] and some may have had their own eyes on

16. On the return journey, which was not part of the trial, the weather was extremely bad, and yet the watch still kept time sufficiently accurately to fall within the most stringent requirement for winning the Longitude Prize.

17. Nevil Maskelyne, the astronomer royal (chief astronomer of England), sat on the Board of Longitude and dragged his feet over awarding the prize to Harrison. Maskelyne was a very capable astronomer; among his achievements was an experiment to measure the

FIGURE 7.9. John Harrison's fifth and most accurate marine chronometer, H5. Figure courtesy of Racklever.

the prize money. After intervention by the king and parliament on Harrison's behalf, he eventually received half the prize money. Meanwhile, he designed and constructed H5 (fig. 7.9), which, trials in 1772 showed, kept time with an error of less than a third of a second per day.

An expertly made copy of H4 accompanied Cook on his second voyage, which lasted three years. The captain expressed himself entirely satisfied with it; his experience demonstrated that the marine chronometer approach to longitude estimation worked not just on sea trials, but on real maritime expeditions.

In recent years John Harrison has received a great deal of attention in the literature for solving the longitude problem. He did so, but we must also give credit to other advocates of the marine chronometer method. Many of the advances in horology during this period came from France (from master horologist Pierre le Roy in particular). Because the Harrison

---

deflection of a plumb line due to the gravitational attraction of a mountain—Schiehallion, in Scotland. Maskelyne was assisted in this experiment by Reuben Burrow, a gifted but pugnacious commoner who, like Harrison, felt that he was not given due credit for his contributions. See, e.g., Danson (2006).

chronometers H4 and H5 were very expensive—they were hand-crafted, idiosyncratic masterpieces—most ship's captains and navigators could not afford to pay for the construction of one of them. It fell to other watch-makers to develop chronometers that were affordable enough to be used by most ships. It was another half century before every ship in the Royal Navy was equipped with a marine chronometer. In the interval, the less-expensive lunar distance method was used. Indeed, many navigators who could afford a chronometer also used the lunar distance method during this period, so that each method could act as a check on the accuracy of the other.

In practice, the early chronometers were handled with kid gloves. They were kept in gimbals in a dry room near the center of a ship to minimize the effects of varying conditions. One of the navigator's key tasks (that's a pun) was winding the chronometer. When making an observation to esti-mate longitude (a celestial altitude measurement, for example), he would use a good pocket watch to record the time, not the marine chronometer, which was never taken outside, so as to avoid exposure to wind or sea. The pocket watch time would be set by reference to the chronometer.

FINDING LONGITUDE BY THE LUNAR DISTANCE METHOD

Advocates of the lunar distance method of ascertaining longitude were doubtful of the efficacy of the rival chronometer method. Newton ob-served that a chronometer, should it go wrong during a long voyage, would render all subsequent estimates of longitude incorrect; the timekeeping error could not be fixed by subsequent observations. The lunar distance measurement, in contrast, was robust in this regard. One bad measure-ment could be ignored because a subsequent good measurement would bring a ship back on course.

The idea of using the moon as a clock to estimate longitude is quite old; it was first proposed in 1514 by Johann Werner. In those days, however, the means to implement the method were lacking. For one thing, the cross-staff instrument was incapable of making measurements with sufficient accuracy; for another, the theory of lunar motion was not developed suffi-ciently to describe the moon's orbit.

How does this method work? The moon's movement across the night sky means that it has a different background of stars at different times. If the orbit is known accurately, an astronomer can predict the (angular) distance between the moon and a specified star and so could say, "At 22:00 hours GMT on April 21, 1750, the moon will be 23° 31′ 10″ distant from Sirius and

GMT

Between the year of its adoption in 1884 and the year it was superseded, 1972, Greenwich Mean Time was the universal clock by which the world ticked. GMT is time referenced to the orbit of the earth around the sun. The earth's variable speed in its orbit, and the inclination of its rotation axis, mean that the apparent motion of the sun around the earth is not constant: it changes speed. Because of this change of speed, a clock that was simply proportional to the apparent position of the sun would tick at a variable rate. To have a solar clock that ticked at a constant rate, GMT was based on an average sun—a yearly average of the sun's apparent movement about the earth. (This averaging is expressed in the "mean" of Greenwich Mean Time.) The averaging process implies that, in practice, there is a difference between actual noon (the time at which the sun is highest in the sky) and 12:00 p.m. GMT of up to 16 minutes.

Universal Time (UT) took over from GMT in 1972, although in England and some other places UT is still called GMT. The difference between UT and GMT is never more than a second. There are several versions of UT. UTC keeps time according to an atomic clock, which is entirely independent of the solar orbit. UT0 takes into account the polar motion of the earth, which is useful for precise geodetic work; and UT1 measures time by the rotation of earth with respect to distant celestial objects called quasars. For more details on the subject of universal time, I refer you to the technical literature—but be warned that it will take you a considerable amount of time to read it.*

* See, for example, Landes (2000), McCarthy (1991), and McCarthy and Seidelmann (2009). NASA also has a website that explains Universal Time.

65° 13' 55″ distant from Draco." (I am making up this simple example— these numbers are fictional.) The lunar distances from a number of prominent stars, against the time at Greenwich when these distances occur, are compiled in an almanac. A mariner observes certain lunar distances, consults his almanac to find the time in Greenwich, estimates local time, and then calculates his longitude from the time difference.

Such is the theory. In practice, the lunar distance method was fraught with difficulties; and at sea, it could not be made as accurate as the chronometer method. Not for lack of trying. The Greenwich observatory had been set up to make astronomical measurements that could be used for compiling navigational almanacs. The sextant had been invented largely to improve the ease and accuracy of the necessary celestial measurements. But making suitably accurate measurements at sea was difficult. Also, many corrections were required to account for the diameter of the moon, atmospheric refraction, and parallax effects. In the days before computers, many hours of mathematical calculation were still required to reduce the observations and obtain an estimate of GMT—and all this measurement and calculation presupposed a clear sky so that moon and stars could be seen. Despite these difficulties, by the end of the eighteenth century, both almanac data and sextant accuracy were good enough to provide a reasonable estimate of longitude at sea, in fair weather.

FINDING LATITUDE AND LONGITUDE BY THE ALTITUDE INTERCEPT METHOD

A sextant was used for celestial navigation, of course, as well as for coastal positioning. Earlier, we applied a sextant to measure bearing angles; here, we turn it 90° and measure altitudes. By the mid-1800s the accuracy of almanacs and of sextants enabled position-fixing anywhere on the globe, at sea or on land, via the *altitude intercept* method. That is to say, the knowledge of star positions in the night sky and the technology of sextant manufacture were both sufficiently well developed for a navigator to determine his latitude and longitude with a couple of celestial readings with his sextant (assuming he had a clear sky and an almanac). Even today, the altitude intercept method is used as a backup to satellite navigation; before GPS, it was a standard technique. This procedure was superior to, and replaced, the older lunar distance method because it did not require shooting the moon: with the lunar distance technique, the angular size of the moon rendered measurements error-prone.

A modern navigator's almanac lists the positions of 57 "navigation stars" for every hour GMT of every day of the year. The chosen stars are bright; and they spread over the northern, equatorial, and southern skies so that, given clear visibility, a navigator will always be able to find a few navigation stars above his horizon. For example, in the northern sky are Capello, Draco, and Vega. In equatorial skies you will find Betelgeuse, Pollux, and Sirius. In the southern hemisphere there are Canopus, Harad, and Rigil Kentaurus. A navigator shoots a few stars with her sextant. She then carries out "sight reduction calculations"—nowadays greatly simplified by software—to correct for observational errors due, for example, to the observer's altitude above sea level or to atmospheric refraction (particularly for stars near the horizon). She consults her almanac to find her stars (this may sound like astrology, but it is solid science). When she finds a star at the observed altitude, she then knows the time at Greenwich. She repeats the procedure for another star to confirm the time estimate. Knowing GMT and the local time at which she made her observations, she can estimate her longitude. Latitude can be found in the same manner or, of course, in the old-fashioned manner from Polaris.

Another way of looking at the altitude intercept method is by considering the geometry. Let us say that our navigator has shot a star and determined its altitude to be the angle a. In figure 7.10a you can see that if the position of the star is known, then angle a limits the navigator's location to a circle, as shown.[18] This circle is the global-scale version of an LoP. Viewed locally it would appear as a straight line, as earlier. Our navigator shoots a second star and determines that her position is limited to a second circle. As shown in figure 7.10b, she now knows that her location on the surface of the earth is at one of the two intersection points of these two circles. Usually these intersection points are widely separated, and so it is easy to choose which one applies. In this case, let us say, our navigator is on a boat in the Pacific; one of the intersection points is in the Pacific Ocean whereas the other is somewhere in western Canada, so deciding which is her true position is not hard. If the two intersection points are close enough together to produce doubt as to which is the true position, then the navigator will shoot a third star and produce a third circle.

Walking the Line: Continental Exploration and Navigation

An English privateer named William Dampier sailed from home to the Pacific Ocean four times during the early 1700s; he circumnavigated the world three times (and produced four best-selling chronicles of his journeys). Sailing the world's oceans was not exactly becoming commonplace or easy, but in Dampier's day it was becoming more common and a little less arduous. The European maritime nations had made the seas of the world their highways, and the traffic was increasing. Sea lanes brought increased trade and spread European wars around the world; the war

18. By "position of the star" I mean its location at the time of observation, where time is GMT. Every star has a position, of course, even if it is below the observer's horizon. Star location in the night sky is specified by two angles: *declination* and *right ascension*, analogous to latitude and longitude.

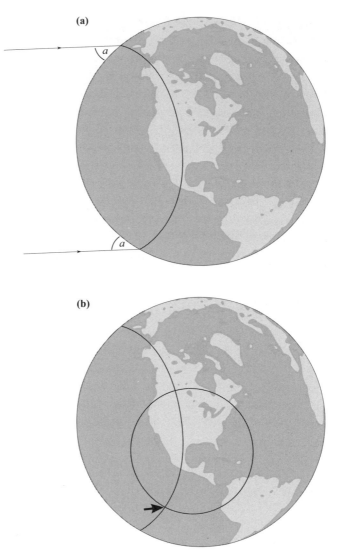

FIGURE 7.10. Celestial navigation by the altitude intercept method. (a) Light from a navigation star arrives at the earth's surface at angle *a*, as measured by an observer. If we know the star's position in the sky, then we know that the observer must be somewhere on the circle shown. (b) A second star observation yields a second circle; the observer is at one of the two intersection points. A third circle would yield an unambiguous location for the observer, but usually, as here, the third circle is hardly necessary.

fought by Dampier's countrymen against Napoleon included a lengthy maritime campaign in the Indian Ocean, for example.

Captain Cook crossed the world three times from England to New Zealand, once traveling westward and twice eastward. A ship's master on one of his journeys, William Bligh, was set adrift near Tonga in the South Pacific after the crew of the *Bounty* mutinied. Bligh and 18 loyal crew members set out westward for Timor in a small open boat with a small sail, six pairs of oars, a compass, and a sextant, and made it there. Bligh and all but one of his crew survived because of their navigation skills; quite generally, improved navigation was easing long journeys (though "easing" is not the apposite word in Bligh's case: his 6,700-km journey was horrific).

Longitude was finally under control, and other aspects of marine navigation were improving. Steam power gradually overtook wind power; a better method of estimating ship speed was introduced (counting the revolutions of a towed rotor); charts were improved, and in particular, hydrographic surveys of the sea bed and harbors became more accurate and complete. At the end of the nineteenth century, radio was added to the maritime navigator's equipment. With all these improvements, the seas were managed, though never mastered or tamed. Sea transport gradually became routine and relatively safe.

Exploration of the world's coastlines was completed before that of the continental interiors. It was not by chance that California was brought into the Union before interior territories, such as Nevada (or that British Columbia was brought into confederation before other western territories of Canada, such as Alberta): travel by sea was surer and faster than travel overland. The epic feats of exploration of the nineteenth century were mostly overland because the means of traveling by land were less developed than the means of traveling on water. The second half of the nineteenth century was the period of colonization by western nations, and numerous expeditions were sent to the unexplored continents of the world—Africa, Australia, and Antarctica.

Many of these expeditions were not truly journeys of discovery because the lands visited were already occupied. They were epic nevertheless, and were seen to be so at the time. The very large Lewis and Clark expedition left the United States at St. Louis in 1804, traveling by river when they could do so, seeking a water route to the Pacific coast. They rowed, rode, and walked a distance of 8,000 miles (13,000 km) before returning in 1806. In the previous decade Alexander Mackenzie reached the Pacific coast from Canada and left his name on a rock by the shore at Bella Coola in

British Columbia. In the decade after Lewis and Clark's journey, David Thompson mapped much of the Pacific Northwest and became the first man to navigate the entire length of the Columbia River. Thompson and Lewis both made celestial observations to estimate the positions of key landmarks along their routes.[19]

South America had been traveled by Europeans since the time of the Conquistadors, but the interior was only really explored much later, by la Condamine, von Humboldt, Ferreira, and others. The Russian expansion eastward into, and exploration of, Siberia took longer and started earlier than the American expansion westward. The vast new territory attracted many explorers, from Semyon Remezov, a cartographer and geographer born in the seventeenth century, to Vladimir Arsenyev, who lived into the twentieth. The coastline of Australia had been studied from the sea well before the interior of that dry continent was explored by Sturt, Leichardt, Burke and Wills, and many others. The first Europeans to explore the interior of Africa did so as late as the Victorian era at the end of the nineteenth century: the expeditions of Mungo Park, Burton and Speke, and Livingston and Stanley were read avidly in the English-speaking world. Much of the European exploration was connected with the carving up of Africa in the scramble for empires during this period.

In sharp contrast with the expeditions to the interior of Africa, in terms of both temperature and purpose, were the expeditions to Antarctica by Amundsen, Shackleton, Scott, Byrd, and others. The other end of the world was more extensively investigated in a search for the Northwest Passage: Parry, John and James Ross, Franklin, Peary, and Amundsen attempted to sail through or trudge across Arctic ice. The story of heroic failures in the far north or south (Franklin, Scott) make for reading that is every bit as inspiring as the successes.[20]

I close this chapter with a land exploration story but not of the regions just mentioned. For its navigational relevance I choose to tell you about the surveying of India, a land which has been occupied since the beginning of history but which was surveyed (with impressive accuracy) in a mammoth undertaking only during the age of sail and steam.

19. The Lewis and Clark expedition has generated many books: see, e.g., Ambrose (2003) and Owen (1979). For more on Thompson and Mackenzie, see, e.g., Hayes (1999) and Warkentin (2007).

20. There is a large literature about land exploration in general, and about particular expeditions, that covers the explorers mentioned here. See, e.g., Fernández-Armesto (2006, chap. 9), Fleming (1998), Hurley et al. (2001), Rice (1990), and Whitfield (1998).

Nain Singh Rawat and the GTS

Following the loss of their American colonies, the British set about obtaining much richer and more populous ones by completing their conquest of India—the country that eventually became the jewel in the crown of the British Empire. In fact, the process was quite piecemeal and was not completed until well into the nineteenth century. The East India Company wanted to survey its new territory to take stock of what it had acquired,[21] but in 1802 it was persuaded to take on a much bigger enterprise called the Great Trigonometric Survey (GTS)—one of the biggest surveying projects ever undertaken and one that would set a high standard for the future. The GTS required a great deal of organization, skill, and money. It took until 1913 to complete and outlasted the East India Company, which was taken over by the crown in 1857.

The GTS surveyed India by triangulation from south to north. The original supervisor, Major William Lambton, may have had higher motives than his employer in undertaking the task: he wanted to make a contribution to knowledge by taking accurate measurements of an arc of longitude (which tells us about the size and shape of the earth) and of gravitational anomalies. His successor, Colonel George Everest, continued the survey northward after taking over in 1818. He introduced the most accurate measuring instruments of the day into the survey and, with his predecessor, measured a large $11.5°$ meridian arc from the southernmost point of the Indian subcontinent to the Himalayas. In 1843 Colonel Andrew Waugh took over the survey, and it was on his watch that the world's highest mountain was discovered; originally the surveyors labeled it Peak XV, but later Waugh named it after his predecessor (who never saw it).

World politics entered the survey at this point. The region that was to be surveyed in the mountainous north of India spilled beyond the border into Tibet, then a closed and fiercely independent society that did not welcome outsiders. Tibet was part of the Great Game between the two main empire-building nations of the world in Asia at that time—Britain and Russia—and the British authorities placed a high importance on staking some sort of claim there by surveying this mysterious land, about which little was

21. The East India Company was a joint-stock company founded at the beginning of the eighteenth century. Its business was trade between Britain and the Far East. It maintained its own army for many years and conquered much of India, which it ran virtually as a private empire until India was formally incorporated into the British Empire by the government in 1857.

FIGURE 7.11. The famous Indian *pundit* Nain Singh Rawat. Nain Singh made three surveying expeditions into Tibet, two of them to Lhasa. Wikipedia.

known in the outside world. To this end, Captain Thomas Montgomerie of the survey trained many native Indians from the northern regions in the disparate arts of surveying and spying. It was thought that Indians would have a better chance of penetrating Tibet than would British surveyors. These *pundits*, as they were called, were taught how to use sextants and record their measurements, as well as how to disguise themselves, travel inconspicuously, and construct cover stories.

Pundits were trained so that they could pace very precisely—exactly 2,000 paces per mile—over all kinds of terrain. In this way they could measure the distances they traveled (yes, they walked the whole way). Disguised as Buddhist pilgrims, they would carry prayer wheels and beads —Buddhist rosaries that were supposed to have 108 beads on them but which had 100 so that the pundits could count paces more easily and surreptitiously. Maps and notes were kept in the wheel, where prayers were supposed to be placed. This was sacrilegious to Buddhists; the pundits risked expulsion, torture, or execution if caught by Tibetan authorities. As each pundit returned, Montgomerie would add the new data to his growing database and use the large corpus of information that he was accumulating (by hook or by crook) to construct detailed maps.

FIGURE 7.12. NASA image of the Himalayas (including Mount Everest) from the north. The pundits had to survey the Himalayas in the 1860s for the Great Trigonometric Survey of India.

Of the many pundits who were employed in this effort, one in particular stands out: Nain Singh Rawat (fig. 7.11). Nain Singh was from the Johaar Valley in northern India, and he and his cousin Mani set out from the survey headquarters at Dehradum for Nepal in 1865. They separated, and Mani traveled through western Tibet, recording his data on the way and then returning to India. Nain Singh walked to the forbidden city of Lhasa, after being robbed of his money and begging for food en route. He met the Dalai Lama but spent most of his time in hiding. By day he recorded the temperature at which water boiled and concluded from his measurements that Lhasa was at an altitude of 3,420 m (close to the actual value of 3,540 m). By night he would emerge from hiding and make celestial observations to estimate his position. After several weeks in the capital, Nain Singh left hurriedly, in fear of his life, and traveled with a caravan for two months before leaving it and returning to India. Disguised as a pilgrim, he had traveled 2,000 km, almost entirely on foot; had made 31 latitude and 33 elevation measurements; and provided a detailed description of the mysterious Tibetan capital.

Nain Singh Rawat undertook a second expedition in 1867, to western Tibet. For his exertions, in 1868 he was recognized by the Royal Geographical Society in London and awarded a gold watch. A third expedition in 1873–75 took him back to Lhasa by a more northerly route than he had adopted on his first journey. In 1877 the Royal Geographic Society bestowed a medal on him, and the governor of India gave him two villages.

Such rewards were very unusual for the Indian pundits. Usually, their wages were low, and they received little in the way of recognition for their accomplishments. Nevertheless, they produced much data that, when brought back to the survey and correlated, led to accurate maps of a mountainous, inaccessible, and previously uncharted part of the world. To appreciate the daunting nature of their task, consider figure 7.12. As one modern-day Indian writer has said, "It is difficult to understand what drove them." Certainly many died, of disease or privation, or were executed, and yet many returned with good data describing their assigned routes.[22]

22. Quotation from Nagendra (1999, p. 14). The Great Trigonometric Survey and the exploits of Nain Singh Rawat and the other pundits are well told in many accounts written over the last 150 years—e.g., Hopkirk (1995), Keay (2000), and Nagendra (1999).

CHAPTER EIGHT

The Electronic Age

Electronic technology has led to a revolution in navigation, in terms of accuracy, scope, and accessibility. In this chapter we see how several electronic navigation technologies work and how they have contributed to the twentieth-century flowering of navigation.

Gyro Navigation

Gyroscopes have been around since the nineteenth century—longer, if you include children's spinning tops. The marine gyroscopic compass (*gyrocompass*) was invented by Dr. Hermann Anschütz-Kaempfe in 1906 and patented in 1908. Elmer Sperry, in the United States, was hot on his heels and patented his own device later the same year. These inventors realized that the complicated physics of gyroscopes could be harnessed to the cause of navigation. Both were motivated by the need for accurate navigation at sea, in an age of increased international tensions and the building up of sea power. Sperry's gyrocompass was quickly adopted by the U.S. Navy, and soon such compasses were on board most ships of the navy and, later, most ships of the world.[1]

The much greater expense of gyrocompasses, as compared with the traditional magnetic compass, is repaid in terms of increased accuracy (typically 0.1–0.2°). The gyrocompass points to true north, not magnetic north. It is independent of the earth's magnetic field and is unaffected by magnetic anomalies, the ship's magnetic field, and suchlike irritations. Another advantage of the gyrocompass is that, being an electromechanical

---

1. Or, to be more accurate, ships and submarines. Anschütz-Kaempfe was motivated to invent a gyrocompass because he was interested in guiding a submarine to the North Pole.

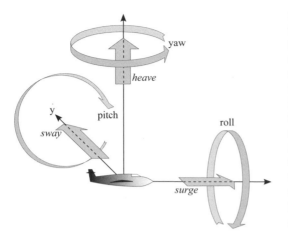

FIGURE 8.1. The six degrees of freedom. An unconstrained moving object can rotate about three mutually perpendicular axes. *Roll* is rotation about the heading direction (here assumed to be horizontal), *pitch* is rotation about the second horizontal axis, and *yaw* is rotation about the vertical axis. There are three corresponding linear accelerations along these axes, known as *surge, sway,* and *heave.*

device, it produces a signal that can be fed into inertial navigation systems and (in a military context) fire control systems. On the other hand, it is a precision-engineered machine—hence its expense—and thus is high-maintenance. Also, it needs a constant source of electrical power.

The gyrocompass is a gyroscope, except that it is motorized so that it never runs down but instead maintains a constant high spin rate—typically about 20,000 rpm (330 Hz). The surprising physics of gyroscopes, based upon precession caused by gravitational torque, ensures that a gyrocompass will try to orient its spin axis so that it aligns with the earth's spin axis—north-south, in other words. The physical explanation of gyrocompasses is often garbled, I have found, so I have made a second exception to my rule about avoiding technicalities and have explained (but still without math) the workings of gyrocompasses in the technical appendix, to which I refer the interested reader.

Gyrocompasses have largely supplanted magnetic compasses as the main shipboard instruments for spatial orientation. (Magnetic compasses remain, as inexpensive backup devices.) Like most other navigational instruments, a gyrocompass is subjected to unwanted forces caused by ship movement. To minimize the effects of unwanted roll, pitch, and yaw torques, it is placed in the central part of a ship; to minimize unwanted surge and sway accelerations, there is some fine-tuning of the instrument. (See fig. 8.1 for an explanation of these terms.) To minimize friction effects, some gyros are suspended in pools of mercury. This arrangement keeps the gyro axis in a horizontal plane, which is important for correct operation (see the appendix for an explanation). It also harks back to the very early days of the first magnetic compass, which, you may recall, consisted of a magnetized needle floating in a bowl of water.

Increasingly, gyrocompasses are being replaced by nonmechanical devices, thus cutting down on precision-engineering production and maintenance costs. There are ring laser gyros and MEMS—of which more shortly—fiber-optic gyros, and fluxgate compasses (electromagnetic versions of the old marine compass, sensitive to the geomagnetic field). All these instruments, being electronic at least in part, are better suited for inertial navigation (IN) applications than the purely mechanical magnetic compass. We will look at inertial navigation in a later section of the chapter.

Gyros are particularly useful in airplanes because of their relatively free movements in all directions. An airplane can rotate (spin) about all three axes and translate (move linearly) along all three axes. It is thus easy for a pilot to become disorientated. The propensity of gyros to maintain a fixed spin axis direction is a great boon for airborne navigators.[2]

RING LASER GYROS

Ring laser gyroscopes are replacing mechanical gyroscopes in many applications because they exhibit a number of important advantages. The physical principle upon which ring laser gyros operate is different from that of mechanical gyros. A ring laser gyro consists of a laser that sends a coherent beam of light around a circular track.[3] The beam is split so that half the light circulates in one direction and half in the other direction. The two beams produce interference patterns. When the ring apparatus is rotating (it is easiest to imagine it rotating about the center of the ring, though the gyro works even if the axis of rotation is outside the ring), the two light beams require different intervals of time to complete a loop, and so the interference pattern changes. The change is proportional to the rotation rate and can be measured with great accuracy. This measurement accuracy makes ring laser gyros very sensitive to rotations. The advantages of this system are significant:

2. This propensity is also employed by firearms manufacturers, who rifle the barrels of their weapons to cause bullets to spin fast. A spinning bullet flies straighter than one with no spin.

3. The track that is followed by the laser light does not have to be circular, but it does need to enclose an area (because the effectiveness of the ring laser gyro increases with the size of the enclosed area). Laser light is made to circulate around a closed track via mirrors or optical fiber.

Spinning Tops and Angular Momentum

The gyroscope phenomenon, familiar to children and scientists, applies to a fast-spinning mass. The mass is usually a disk weighted more heavily on the outer rim than it is near the center. The mass must have an axially symmetric shape, meaning that its appearance is unchanged if you rotate it on its axis. Such an object, when set spinning rapidly, is resistant to changes in its orientation. That is, a gyroscope will resist forces that try to change the direction of the spin axis. We see this in the resistance of a spinning gyroscope to falling over, whereas the same gyro does fall over when it is not spinning. The strange and enchanting aspect of the gyroscopic phenomenon is this: a gyro subjected to a force will move, not in the direction of the applied force, but in a direction that is perpendicular to it. This is very odd and counterintuitive behavior.

The figure shows a familiar example of a gyroscope: a toy top spinning on a table. The quality of this top that is responsible for the odd behavior is *angular momentum*. Unfamiliar? Think of ordinary (linear) momentum—the resistance offered by a moving body to changes of speed or direction. It is easier to catch a baseball than a cannonball moving at the same speed because the cannonball has more momentum and will resist being stopped more strongly. Angular momentum is the rotational equivalent: an object

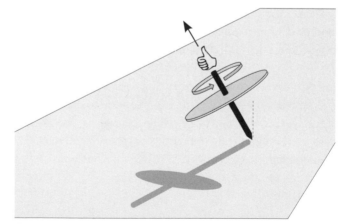

A toy top. The top's direction of spin is determined by the right-hand rule: fingers curl around the rotation axis in the same way as the spin, and the extended thumb gives the spin direction. In this case the spin—and the angular momentum vector of the top—is directed upward, as indicated by the arrow. A fast-spinning top will resist forces that try to change this direction.

with a lot of it resists changing its spin speed or direction. But what exactly is the "direction" of a spinning object? In the illustration you see how physicists assign a direction to spin. The top shown is spinning counterclockwise, seen from above. Curl the fingers of your right hand about the axis, in the same counterclockwise sense—your extended thumb points in the spin direction. For a clockwise spin, the direction is into the table.

A top has high angular momentum because it is spinning fast, and so it resists forces that try to change its spin direction. This is why a spinning top will not fall over, but one that is not spinning does fall. What about the odd phenomenon whereby a top moves perpendicular to an applied force? This is harder to explain; see the appendix if you need to know. It is a consequence of angular momentum, and it underlies the workings of a gyrocompass.

- A ring laser gyro has no moving parts (except the light).
- It is small, lightweight, and robust.
- It does not resist changes in orientation because it depends on a principle of optics and not on angular momentum.[4]

These characteristics mean that ring laser gyros find applications where mechanical gyros cannot be used—for example, on board a missile or an airplane that is maneuvering wildly.

Because a ring laser gyro is sensitive only to rotations in the plane of the ring, a navigation system must include three ring laser gyros oriented perpendicular to each other, to enable a navigator to measure rotations about all three spatial axes. Such an arrangement is employed in modern inertial navigation systems.

MICROELECTROMECHANICAL SYSTEMS (MEMS)

Microelectromechanical systems, or MEMS, have evolved over the last 20 years as an offshoot of the already-mature microchip fabrication industry (fig. 8.2). These little machines (less than 1 mm across) are electrically powered mechanical devices and find all kinds of applications. Of interest to us are the applications as miniature gyroscopes and as accelerometers. The physical principles upon which MEMS act are different

4. The optics principle underlying ring laser gyros, first understood in 1913, is called the *Sagnac effect*. It is a relativistic phenomenon, which is to say its full explanation requires an understanding of Einstein's theory.

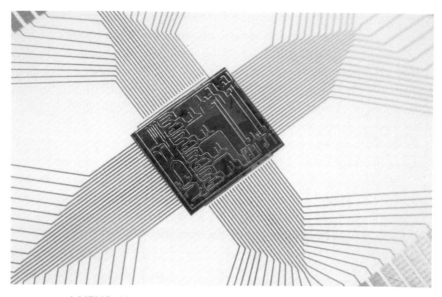

FIGURE 8.2. A MEMS chip. Low power, small size, and low cost have led to many and varied applications of MEMS technology, including acceleration and rotation sensors in navigation instruments. Photo by Maggie Bartlett.

from those of mechanical gyrocompasses and ring laser gyros. MEMS devices work by converting the mechanical movement of (small) parts within the device into an electrical signal using the piezoelectric properties of the MEMS components or else their electrical capacitance. In the first case, we are exploiting a characteristic of some crystals (both natural crystals such as quartz and man-made crystals) to generate an electric charge when mechanically stressed. In the second case, the capacitance of two small blocks of substrate material depends upon their physical separation, and this separation changes when the substrate is accelerated or rotated.

MEMS thus consist of a small mechanical sensor and a microprocessor that monitors, processes, and outputs an electronic signal indicating the existing levels of torque or acceleration. Currently, MEMS devices are not as accurate as macroscopic measuring instruments, but they already have many applications. MEMS gyros are employed in some automobiles to sense yaw: if a threshold is exceeded, stability control mechanisms kick in. MEMS accelerometers (accelerometers that measure linear acceleration) are used in electronic games controllers, airbag deployment, inkjet printers, and a thousand other devices. MEMS are inexpensive, small, and robust; and they consume little power.[5]

You can see that three mutually perpendicular gyrocompasses and three mutually perpendicular accelerometers can monitor all six torques and linear forces that act upon a moving object like an airplane, a ship, or an automobile (see fig. 8.1). Such a collection of gyros and accelerometers form the basis for an inertial navigation system. These systems have been a mainstay of modern navigation for a few decades, and now it is time for us to investigate them.

Inertial Navigation

Just as technology and an increased understanding of physics led to improved compasses, so they also led to an improved version of the old dead reckoning navigation technique. Inertial navigation (IN) is an automated version of dead reckoning, with more accurate and much more frequent updates to estimated platform speed and direction. Here is how it works.

In addition to accelerometer and gyro sensors, the third component of an IN system is a computer. The computer frequently monitors the sensors' output (many times per second in the case of an airplane IN system) and uses these data to continuously calculate the position, speed, and heading of an airplane (or missile or ship). Inertial navigation works without reference to the outside world; all the data that are gathered refer to the accelerations and rotations of the platform itself. This feature is both an advantage and a disadvantage. In the context of military platforms—be they nuclear submarines, fighter planes, or guided missiles—the fact that an IN system works independently of external data means that it cannot be electronically jammed. On the other hand, it means that the IN system can drift: as with our dead reckoning example earlier, IN errors accumulate. Therefore, eventually, it is necessary to update and calibrate the IN track with some sort of external data. GPS is commonly used for this purpose these days. Modern IN systems can maintain a platform bearing that is within 1° of the desired course for one hour in the absence of external position updates. This is plenty of time for landing an aircraft in conditions of poor visibility, or for the system to serve temporarily as an automatic pilot.

In general, IN systems are too expensive for most civilian applications.

5. To learn more about MEMS, see Acar and Shkel (2008, chap. 5).

They arose in World War II to guide German rockets to their target cities and have developed markedly since then. By the late 1950s, IN helped guide a U.S. submarine beneath the polar ice cap. The developments in IN sophistication have been theoretical as well as material. That is, engineers' understanding of how best to process the input data of an IN system has developed in tandem with improvements in accelerometer and gyro accuracy and robustness. Algorithms such as Kalman filters integrate external and internal data in a manner that has been shown to minimize track error. More than that: it has been shown that when such filters are applied to IN data, the resulting errors are stable—they do not grow without limit as they did for the old-fashioned dead reckoning methods.

The development of IN systems over the past 60 years—a pointer to the future—has seen them evolve from large electromechanical instruments held in low-friction gimbals (so that they are insulated from unwanted shocks and vibrations) to small electronic or electro-optic devices that are much more robust and so can be strapped down to a platform—any platform. The robustness and small size of modern IN components have greatly increased their applicability. A trend of the past 20 years is likely to continue: the increasing use of external data and the integration of many different forms of remote-sensing data (GPS, radar, radio altimeter, infrared, sonar, and so on). Inertial navigation devices and other navigation instruments are merging into larger automatic navigation units, with an increasing number of sources of sensor data input and therefore an increasing capability for autonomous operation in many and variable environments.[6]

Radio Direction Finding (RDF)

Radio direction finding is almost as old as radio itself. Radio waves were discovered in the last decades of the nineteenth century. They have a much longer wavelength than other kinds of electromagnetic waves, such as microwaves, light waves, infrared radiation, X-rays, and gamma rays. Whereas visible light has wavelengths of about 500 nanometers (nm)—so that 2,000 waves would extend a distance of 1 mm—radio waves have wavelengths that vary from somewhat less than a meter to hundreds of kilometers. The utility of radio waves is that they can convey information as a result of amplitude modulation (AM) or frequency modulation (FM). Radio waves

6. For a good technical (though nonmathematical) review of the first 40 years of IN development, see King (1998).

are attenuated—scattered or absorbed—by the earth's atmosphere much less than are most other electromagnetic waves, so they can travel longer distances. Hence, from the beginning, they have been used for communication between widely separated people without any physical connection between them, such as a telegraph line. Such *wireless telegraphy* is obviously a great convenience because it means that ships at sea, airplanes, and other mobile or remote platforms can send or receive information without trailing communications lines behind them as they move.

AIR WAVES

The first radio communications patents were granted at the end of the nineteenth century. Our understanding of radio wave generation, propagation through the atmosphere, and reception increased enormously over the first few decades as the usefulness of the new technology was increasingly exploited. The first RDF application dates back to 1902, and the technology was maturing rapidly by World War II, to the extent that radio communication and navigation were crucial components of warfare and navigation during that violent time. Among the radio wave phenomena that were understood during this period was over-the-horizon (OTH) propagation—the ability of long-wavelength radio signals to travel in curved paths. This is different from light-wave propagation. Light travels in straight lines, with only a slight deviation due to refraction through the atmosphere. Short-wavelength (i.e., high-frequency) radio waves do likewise, but long-wavelength (low-frequency) radio waves follow the curve of the earth's surface, so that long-wave radio receivers can pick up a signal from a transmitter that is over the horizon—below the line of sight. Indeed, such radio signals can travel thousands of miles. To understand this phenomenon—and it is an important consideration for radio navigation—we need to look at the atmosphere from the perspective of a radio engineer.

A wave—any wave at all, not just a radio wave—can deviate from a straight line in one of three ways: reflection, refraction, and diffraction. A sound wave bouncing off a cliff face causes an echo; a light wave bounces off a mirror. These are familiar examples of *reflection*. The sun appears flattened just before it sets because light traveling a long distance through the atmosphere is bent downward slightly; water waves obliquely approaching a shelving beach line up parallel to the shore and thus change direction. These are familiar examples of *refraction*. The same water waves passing through a narrow harbor entrance will fan out; light passing through a pinhole lens similarly fans out. These are examples of *diffraction*. The atmosphere gives rise to refraction of radio waves, which we can under-stand by considering how the atmosphere appears (to a radio engineer) to be layered.

The bottom 12 km of the earth's atmosphere is called the *troposphere*. All our weather—cloud formation, pressure and temperature variations, pre-cipitation levels and humidity fluctuations—happens within this layer. These meteorological effects influence the attenuation of radio waves and thus the range of radio communication. Electrical storms within this layer interfere with radio transmission and reception. Above the troposphere, extending from 12 km up to an altitude of 50 km, is the *stratosphere*. Here, the air density is pretty constant, and water levels are low. Radio waves travel in straight lines through this layer, and apart from a slight attenuation—a small reduction of wave intensity with each kilometer the waves pass through—the stratosphere is like the vacuum of space so far as radio-wave propagation is concerned.

Above the stratosphere lies the *ionosphere*. This layer is thick (extending upward from 50 km for hundreds of kilometers to the edge of space), but the air in it is very thin. Despite its low density, the ionosphere has a much greater influence on radio wave propagation than the much more substan-tial layers beneath it. Solar radiation ionizes the molecules of air in the ionosphere, altering the atmospheric electrical conductivity and causing radio waves that reach it at a shallow angle (say, those that were trans-mitted horizontally from a ground station) to refract. These waves follow a curved path that returns them to the lower layers of the atmosphere. Such waves may reflect off the earth's surface and bounce between surface and ionosphere, or they may refract off the lower levels of the ionosphere, pass through the stratosphere, and refract again off the ionosphere. In short, the surface and the ionosphere act like a waveguide, channeling radio waves along the gap between them, as illustrated in figure 8.3. This phenomenon leads to OTH radio; such trapped waves can bounce several times off the ionosphere and travel a long way around the world.[7]

The radio waves that follow the curvature of the earth's surface because of ionospheric refraction are called *ground waves*. Many of the earliest radio and RDF applications used ground waves. These waves must have long wavelengths because shorter-wavelength radio waves are attenuated more strongly by the troposphere and do not refract so markedly through the

7. OTH radar also exists: long-wave radar signals are bounced off the ionosphere to detect targets that are over the horizon and out of sight.

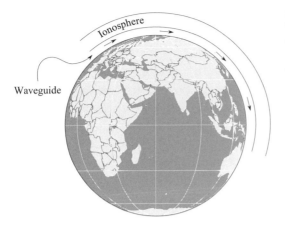

Waveguide

Ionosphere

FIGURE 8.3. Ground waves. The ionosphere traps low-frequency radio waves that are generated by terrestrial transmitters; they can travel over the horizon for thousands of kilometers.

ionosphere. Short waves travel in more or less straight lines, as does visible light; such radio waves are called *sky waves* by engineers and radio hams. The downlinks from GPS to surface stations are via sky waves, and the refraction effects must be understood so that their influence on propagation time from satellite to receiver—crucial to GPS operation, you may recall—can be anticipated. Given the complicated physics of the ionosphere, radio engineers have divided the ionosphere into several sublayers, based on their differing properties. The intensity of these layers (that is, the refractive effects upon radio waves) varies with the time of day, and disappears or is lessened at night, because there is no sunlight to ionize the air molecules. This varying intensity of ionization means that the propagation characteristics of the ionosphere are much different at night than they are during daylight hours. We do not need to look into these additional complications here; ionospheric refraction and the consequent OTH phenomenon of ground waves are all we need to know for an appreciation of RDF and radio navigation.

NONDIRECTIONAL BEACONS (NDBs)

If you don't like acronyms, then the subject of navigation in the electronic age will cause you boundless irritation: it is full of TLAs (three-letter acronyms). In this section we investigate nondirectional beacons, which constitute the earliest contribution of radio technology to navigation. An NDB is simply a radio transmitter that sends its signal in all directions equally. It can be used for navigation in exactly the same way that a star is utilized to obtain an LoP. The position of the transmitter must be known: it will appear on the navigator's charts, and of course, the navigator must have a radio receiver that is sensitive to the transmitted frequency. Given this simple equipment, accurate LoPs were obtained in the early decades of the radio age via NDBs for ships and airplanes. If the radio waves were long, so that they propagated over the horizon, the LoPs traveled along great circle routes, so a navigator had to allow for this when he drew the LoPs on his chart. Of course, nowadays computer software can take care of such details and draw the LoPs on an electronic map on a display screen for whatever map projection is used. For shorter ranges—say, of a higher-frequency radio beacon at a harbor entrance transmitting line-of-sight and received by an approaching ship—such earth curvature effects are unimportant.

Several LoPs drawn from different NDBs, all at known locations marked on a navigator's chart, permit a position fix. In a military context, the advantage of the old NDBs was that they did not give away any information about the location of the navigator because their signal could be received from any direction relative to the transmitter. All of the direction-finding information was gleaned locally by the navigator operating the receiver. These early RDF devices were better than older methods of obtaining a position fix because they were easy to use, as we are about to see, and they worked in most weather conditions and during hours of darkness. A 2-kW transmitter could be detected from a distance of 140 km (75 nautical miles).

Both specially constructed towers and existing lighthouses supported NDB transmitters to aid inland and coastal navigation. Each would have its own unique frequency or Morse code identifier. The Cunard Line of British shipping was the first to use radio beacons for navigation, in 1911.[8] Thirteen years later, airborne navigators were employing RDF. In the United States, commercial radio stations acted as NDBs, each transmitting its unique radio station identifier once per hour for the benefit of airborne or marine navigators. Many NDB transmitter networks were constructed around the world in the 1920s and 1930s, particularly along coastlines; a few still exist.

One disadvantage of early NDBs is that they were not particularly accurate. I can illustrate why with a simple made-up example of a RDF receiver.

In figure 8.4a we have a radio receiver that consists of three antennas arranged on a wheel. The average power received from the NDB transmitter

8. The earliest systems placed the direction-finding receivers on shore. An incoming ship would transmit a signal and its direction to a transmitter—say, at a harbor entrance—and the coastal receiver station would then radio back to the ship.

via these three antennas depends on the orientation angle, a, of the antenna wheel relative to the transmitter direction. The dependence of receiver power on wheel angle a is shown in figure 8.4b. There is a peak in the receiver output signal at zero angle when the wheel is oriented with antennas 1 and 2 equidistant from the transmitter. Thus, my simple receiver—let us call it a *Mercedes receiver*—acts as an RDF: it finds the direction of an NDB transmitter and so provides a navigator with an LoP.

But also note the width of this Mercedes receiver beam shown in figure 8.4b. Such widths are conventionally measured in terms of the angular extent of the beam at 50% of peak power; in this case the beam is about 50° wide. In principle, a navigator could manually adjust the receiver wheel of figure 8.4a until it pointed in the exact direction of peak power, but in practice, there would always be an error: it is not possible to estimate the peak angle exactly. Radio engineers found that it was easier to precisely detect *nulls* (angles for which they received zero power) than peaks. As a result, real antennas were designed so that, unlike the example of figure 8.4, the receiver emitted zero signal only when it was aligned with the transmitter.

Another problem inherent in the design of my Mercedes RDF receiver is seen if we extend the beam pattern of figure 8.4b—the graph of average power received versus orientation angle—to larger angles. The same peak is displayed at angles of 120° and 240°, so a navigator who used this type of receiver would have to derive some method for deciding which of three directions that produce an output peak corresponded to the true direction of the transmitter.

A very common and simple receiver antenna used in real systems (unlike my hypothetical Mercedes receiver) was the *loop antenna*. It produced a null output signal if the plane of the loop was perpendicular to the direction of the transmitter. This design meant that a navigator had to choose one of two nulls separated by 180° (instead of three peaks separated by 120°, as for the Mercedes design). Supplementary circuitry sorted out this ambiguity.[9] Because loop antennas could be small, these anten-

9. Despite the relative ease of sorting out the 180° ambiguity of transmitter direction—which was due to the symmetry of the loop antennas (they looked the same from the back as from the front)—there were still serious navigational accidents that resulted from misidentifying the true direction of the transmitter. In 1923 a squadron of 23 U.S. destroyers ran aground on the coast of southern California because of this elementary error, and 6 ships were lost. See Cutter (2004) for this detail and for a summary of early RDF developments in marine navigation. RDF is still practiced as a hobby by radio enthusiasts around the world.

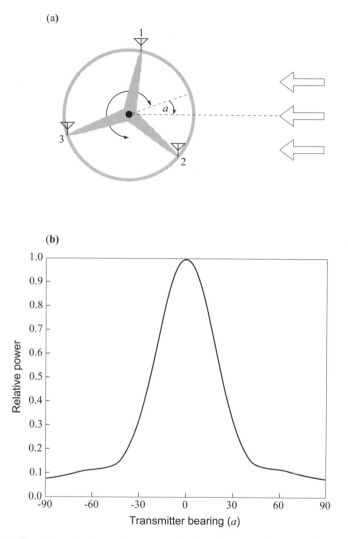

FIGURE 8.4. The "Mercedes" RDF receiver. (a) Waves from a radio transmitter (arrows) are picked up by three receivers distributed around a wheel. The direction of the transmitter relative to the wheel is given by bearing angle *a*. (b) The sum of the signals received by the three receivers depends upon transmitter bearing angle *a*. In this graph the power of the signal sum is plotted versus *a*. The transmitter direction can be found by rotating the wheel until the peak signal is obtained.

nas were employed for RDF aboard aircraft. As expected, the nulls were sharper (extending over a narrower range of angles) than the peaks, so navigators preferred to use nulls for obtaining an LoP from a loop antenna.

LORENZ SYSTEMS

You may think that you have escaped from TLAs, at least for one section of this chapter, given the name of this German RDF system developed during World War II (though *Lorenz* was the name given to it by the Allies). You would be wrong: Lorenz is one of many examples of LFR (low-frequency receiver) devices developed during the 1930s and 1940s, in which radio direction finding was achieved by transferring part of the task to the transmitter. That is, instead of the receiver and the navigator performing all the work of estimating transmitter direction, as with my Mercedes receiver in figure 8.4, a significant part of the task was carried out by modifying the transmitted signal in some way. This trend would continue into later decades, when all of the direction-estimating work would be carried out by the transmitter. This eased the navigator's task, as we will see.

The Lorenz system was used by German bombers to guide them toward British cities during the later stages of the Battle of Britain. It worked as shown in figure 8.5. Two narrow-beam transmitters—each with a beamwidth of only a few degrees—overlapped a little. Each beam transmitted a modulated signal, but (here is the clever part) the signals in the two beams were out of phase. The very slight overlap between them was pointed in the direction of the target city. An aircraft that flew in one beam or the other would detect a modulated signal, whereas one that flew in the overlap region would detect a constant signal. German aircraft would fly along the beams, and their navigators would know that they were on the right bearing when their Lorenz receivers produced a constant signal. If the received signal was modulated, they were too far to the left or right and would make a course correction.

Note how the onus of direction estimation is moving from the receiver to the transmitter. The complicated equipment was back at base, on friendly soil; as a consequence, the airborne receivers were small and simple. The Lorenz system was successful for a while, until British electronic countermeasures defeated it.[10]

10. Actually, the Lorenz transmitters were in France, which was not so friendly to the Germans; but at that period the French were not overtly hostile, having been conquered a few months previously. The Battle of Britain saw the first large-scale adoption of electronic warfare and electronic countermeasures, aimed at guiding German bombers to their target

VHF OMNI RADIO (VOR)

There were two trends in evidence as radio direction finding—RDF—evolved over the decades of the twentieth century. First, as we have seen, the task of estimating direction moved from receivers to transmitters, reducing the complexity and thus the cost and size of receivers and increasing their portability. Second, the transmitter frequencies tended to increase, so that later systems generally transmitted a medium-wave signal instead of a long-wave signal.[11] Shorter wavelength means that a transmitter or receiver of given size can form a sharper beam—a narrower beamwidth. There are a number of advantages to a sharper beam. Power is concentrated where it is needed, so the system has a longer range; in a military context the signal is less susceptible to electronic jamming; for navigators it means that directional beams are more accurate.

VOR is a high-frequency RDF; the letters stand for "VHF omni radio," with VHF meaning "very high frequency" (didn't I warn you about TLAs?) "Omni" means omnidirectional: in the older NDBs the transmitter sprayed out a signal equally in all directions. Yet VOR was unlike the NDBs because the signal that was transmitted depended upon direction. The transmitter rotated in sync with a modulated carrier signal so that the phase of the transmitted signal varied in a predictable way with bearing angle. An aircraft with a VOR receiver could decipher the phase that it was picking up and thus know the direction to the transmitter. Furthermore, the VOR receiver did not have to be rotated at all; the moving parts were all in the transmitter.

VOR became the standard RDF method for civilian aircraft navigation during the 1950s and 1960s. Most countries adopted a two- (or even three-) component system, each with its own receivers, and each operating over a restricted range of aircraft altitudes. Hence, an airliner approaching a runway would pass from one system to another as it descended. The two or three VOR systems for civilian aircraft guidance had different ranges and accuracy requirements, and they operated at different frequencies so that they did not interfere with one another.

---

cities—or misdirecting them away from these cities. The original Lorenz had a short range and was superseded by more powerful and sophisticated aircraft navigation systems that employed the same principle. See Denny (2007) for a summary.

11. Long-wave radio covers the bandwidth 150–400 kHz, whereas medium-wave radio covers 520–1720 kHz. Thus LW signals can have wavelengths as long as 2 km whereas MW signals can have wavelengths as short as 175 m.

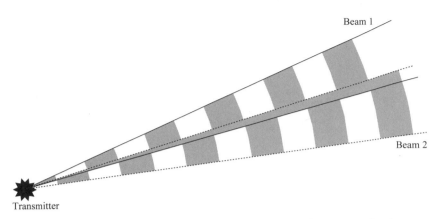

FIGURE 8.5. The Lorenz system. Bomber aircraft were directed to a target city via two transmitted signals of narrow beamwidth. Here beam 1 is delineated by solid lines, and beam 2 by dashed lines. The two transmitters were located close to each other, and the beams overlapped slightly. The transmitted signals were modulated 180° out of phase so that in the overlap region, and only in this region, the receiver signal was unmodulated. An airplane with a Lorenz receiver that produced an unmodulated signal would therefore have been within the overlap region.

VOR systems operated by line of sight (they did not benefit from the OTH capability of longer-wave systems, and anyway, they were generally intended for short-range applications). They were more accurate than NDB systems: a VOR receiver could estimate the direction of a transmitter to within 1°. Modern versions (VOR has evolved considerably over the decades) are less susceptible to intentional or natural sources of electronic interference. VOR is now being phased out (another pun, I fear) in favor of GPS navigational systems. (We saw in chapter 3 that the FAA was prominent among advocates of civilian use of undegraded GPS transmissions.)

LORAN SYSTEMS

At last—an RDF system with a name that is not a TLA. In fact, the *Long Range Navigation* system is also exceptional in another way: it bucked the gradual trend in RDF of increasing transmitter frequencies by lowering its frequency over the decades as it matured. By choosing a wavelength of 3 km, LORAN-C transmitters generate ground waves that are capable of traveling 5,000 km—well over the horizon.

LORAN is an example of a *hyperbolic* navigation system. A forerunner of GPS in some ways, LORAN was based on timing differences between widely separated transmitters. The idea is shown in figure 8.6. Transmitters

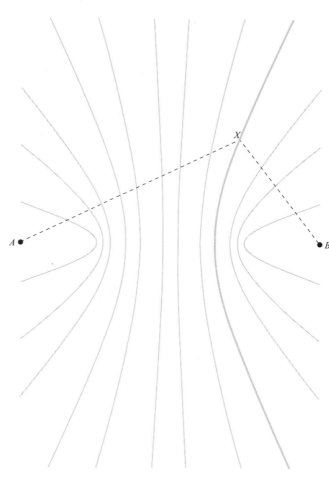

FIGURE 8.6. LORAN—Long Range Navigation. Two signals transmitted simultaneously from A and B will reach the receiver, X, at different times, unless the receiver
happens to be equidistant from the two transmitters. From the time difference we
know that the receiver must lie somewhere on the bold LoP—a hyperbola. (For a different time difference, the receiver would lie on one of the other LoPs shown.) Stated
differently, the length AX differs from the length BX by the same amount anywhere
along the curve. To fix the receiver location, a third transmitter signal must be picked
up by the receiver. This third signal yields another hyperbolic LoP, which intersects
the bold line, fixing the receiver's position.

A and *B* send out signals modulated in a manner that permits a receiver to estimate the time difference between them. From the detected time difference, a navigator can construct an LoP: it is a hyperbola.[12] (Several hyperbolas are shown in fig. 8.6, corresponding to various time differences.) Consequently, an LoP derived from LORAN was known as a TD line (time-difference line). A third transmitter leads to a second LoP; the intersection of these two hyperbolic LoPs yields a position fix. This method of determining position is not triangulation but instead is a two-dimensional version of the *trilateration* technique later used by GPS.

The first LORAN system was developed in World War II from a British system for guiding bombers onto German cities. LORAN—an American development, more powerful and sophisticated than its predecessor—had a longer range; and eventually British as well as American bombers used LORAN to help them navigate to their target areas. Unlike the German Lorenz beam rider method, this hyperbolic system did not reveal to the enemy the probable target city of the bomber planes. This first version transmitted signals in the frequency range of 1.8 to 2.0 MHz. LORAN-A lasted well beyond the war years. Its transmitter network was extended, and the system became very popular among commercial fishing fleets and other marine vessels because it worked quite well in most weather conditions and because a LORAN-A receiver was inexpensive.

The LORAN-A transmitters were maintained until 1980, but long before then, an improved version, LORAN-C, had been introduced. This version operated at lower frequency (around 100 kHz) and consequently was able to see over the horizon. Many hundreds of LORAN-C transmitter towers were constructed around the world, some of them very powerful (fig. 8.7), with ranges exceeding 5,000 km. LORAN-C was the mainstay of airline navigation for four decades, from about 1970 to 2010, when the U.S. transmitters were switched off. It was a reliable system that provided navigation fixes for airplanes or ships anywhere within the coverage area. The coverage was not universal, however, and the system required expensive infrastructure. Moreover, it was susceptible to ionospheric fluctuations at dusk and dawn, or to fluctuations in atmospheric ionization caused by

12. Because the speed of light (and of a radio signal) is constant, a measured time difference converts into a well-defined difference in path length from the receiver to the two transmitters. Mathematics tells us that the receiver must lie somewhere on a hyperbolic line passing between the transmitters; which hyperbola depends upon the size of the path length difference.

FIGURE 8.7. The LORAN-C transmitter on Marcus Island, off the coast of Japan. When constructed in 1964, the tower was 412 m high, and the transmitter power was 4 MW. In 1985 the tower was replaced by one half as high and transmitting 1.1 MW. There were hundreds of LORAN-C towers around the world during the four decades that this system guided international air traffic. U.S. Navy image.

solar flares. The accuracy of LORAN-C depended upon receiver range to the transmitters. It could be used to estimate position to within 15 m if all three towers were within 300 km, but if the transmitters were 1,500 km away, accuracy was degraded to about 500 m.[13]

Radar

Like its close relative LORAN, radar—an acronym for "*radio detecting and ranging*"—emerged from World War II as a mature technology, having been developed rapidly during the war years. In fact, the radar technique for spotting incoming airplanes had been hastily brought into service in England immediately before the start of the war and proved itself during the Battle of Britain. Radar would go on to prove decisive in another

13. For an easy-to-read nontechnical account of LORAN-C use, see Stearns (1980). A technical summary is provided by Clausing (2007, pp. 156–65). There was an intermediary version of this navigational network known as LORAN-B, never deployed widely, as well as a military version, LORAN-D.

campaign, by providing the Allies with the capability to detect U-boat periscopes from the air during the long and hard-fought Battle of the Atlantic. By the end of the war, more effort and expense had been put into radar development than into the development of any other technology apart from nuclear bombs.

The rapid maturing of radar as an independent discipline can be seen in the many different applications of radar equipment evident today. Most of these were known, in theory if not in practice, to the early wartime developers. Nowadays radar is a crucial component of weather prediction, terrain mapping, airplane altimetry and air traffic control, and space missions. Of course, it still has many military applications, such as missile guidance and gun-laying, airplane and missile tracking, and air and sea surveillance. A spin-off of radar proper, medical imaging uses remote sensing technology and algorithms that are taken directly from (or which follow on from) radar remote sensing.

Several of these radar fields lend themselves to navigation applications. The radar equipment and processing techniques used in navigation vary considerably from one application to the next. The shape and size of radar antennas (both transmitters and receivers); the transmitted frequencies; the shape, number, and power of transmitted radar pulses; the detection range; frequency bandwidth; and display screen attributes—all differ significantly between radar altimeters and synthetic aperture mapping radars, or between short-range missile trackers and Doppler weather radars, or between surveillance radars that are designed to look up at airborne targets and those that are searching for terrestrial targets.[14]

I concentrate here on the navigation applications of radar and provide only a brief summary of radar operation, making no further reference to the techniques used to process signals or to the wider implications of our expanding knowledge of this important field. Radar surveillance, tracking, and mapping impinge on modern electronic navigation capabilities. Arguably, the most important facet of radar for navigators today and in the future lies with the sensor fusion potential of different electronic remote-sensing techniques; data obtained from different sources can be displayed

14. In radar terminology, a *target* is simply the object that a radar operator is looking for, whether or not he intends to shoot at it. *Clutter* is everything else that he sees on his radar display. Thus, rain and other precipitation are clutter to a naval radar operator who is trying to detect incoming missiles, but these atmospheric phenomena are a target for a meteorologist who is using radar to gather data for weather forecasts.

upon a radar screen, as we will soon see, thus providing navigators with a lot of information in a form that they can rapidly assimilate and act upon.[15]

BLIPS AND BEAMS

Radar is a remote sensor, and its operation differs significantly in one respect from that of the other radio techniques that we have investigated thus far. Radar equipment does not passively detect what is out there: it transmits electronic signals into the outside world and then listens for the echoes. This two-way process means that radar operation does not depend upon a cooperative target. A civilian airliner transmits signals to aid in its detection by an airport air traffic control (ATC) radar, whereas an enemy fighter plane does not want to be detected, yet both may be seen by radar. The basic principle of operation is simple. It depends upon directing a radar signal and timing its echo. That is, a radar transmitter sends out a signal (for simplicity, let us say that it is a single, brief "blip") in the direction in which the transmitter antenna is pointing (let us say that the beam is very narrow, so that the direction is well defined). A clock starts when the blip is sent out and stops when its echo is received. Knowing the speed of light—which is also the speed of radar microwaves—we can calculate the target range. Thus the simplest radar set provides range and bearing for an object of interest, a target.

To increase coverage area, the transmitter beam is rotated, so that, for example, a beam sweeps 360° around the horizon. The idea is sketched in figure 8.8, where we see a terrestrial radar antenna beam sweeping over the sea. Surface and airborne targets are *painted* (illuminated by the beam). A transmitted blip is bounced off each and is reflected back toward the radar, where a receiver antenna picks up the signal, processes it, and displays the data on a screen, such as that shown in figure 8.9.

Between the simple diagram of figure 8.8 and the display of figure 8.9 lie 70 years of accumulated knowledge about how to filter, amplify, and process radar signals. First the basics: we have seen how target bearing and range can be determined through use of a directed beam and timing. It is a straightforward matter to display this information on a radar screen that shows a plan view of the radar coverage area. The screen shown in figure 8.9 has such a display, with range circles centered on the radar location

15. I worked for 20 years in the field of radar signal processing. To my surprise, I found that there were remarkably few popular-level books that sought to explain radar operation in a nontrivial way, but without resorting to the heavy mathematics of dedicated textbooks, and so I wrote one: see Denny (2007).

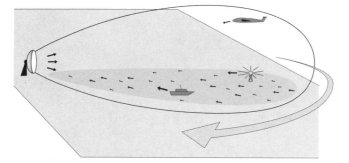

FIGURE 8.8. Radar basics. A transmitter beam sweeps over an area of sea, "painting" the surface and the air above it. A blip, or pulse, of microwave energy is transmitted along the beam; it reflects off all it encounters, and the reflected signal is picked up by the radar receiver. Transmitted and reflected signals are here represented by arrows. Signal processing enables the radar operator to discern useful targets—such as a boat, a beacon, or an airplane—from clutter, such as the sea surface. Further processing may yield the altitude of the airplane and the speed and heading of the plane and boat. Even the most basic navigation radars can provide a reliable estimate of a target range and bearing.

overlaid on the display, for easy reference. But how do we tell whether the target is on the surface of the sea or above it? A plan view will show the latitude and longitude of a target but not its height above the surface. More importantly, the radar beam of figure 8.8 cannot distinguish between surface targets and those at altitude; all are painted by each blip within the beam. Another basic problem is the clutter: our transmitted blip will reflect off the sea surface as well as off the intended targets—and the sea is much bigger than the targets, with a correspondingly overwhelming radar signal.

In fact, all of these problems can be overcome. The radar beam can be jittered[16] and, with appropriate processing, the airplane altitude extracted. This altitude can then be appended as a number beside the signal plotted on the radar display. Because sea clutter fluctuates in a characteristic manner, its effects can be mitigated by filtering and signal integration. The point is that modern radar systems are extremely capable as a result of the advanced signal processing that manipulates the raw receiver signal data. Target speed can be determined with Doppler radar. Many targets can be tracked simultaneously on a display, as for the ATC radar of figure 8.9. Another type of radar system—synthetic aperture radar, or SAR—exploits

16. "Jitter" in this context means rapidly oscillating the radar beam; the resulting data can be processed to provide an improved target position estimate.

FIGURE 8.9. Air traffic control radar aboard the aircraft carrier USS *Ronald Reagan*. Note the range circles superimposed upon the display. U.S. Navy photo by Mass Communication Specialist Seaman Mikesa R. Ponder.

moving airborne platforms to generate terrain maps of increasingly high resolution. In recent years, new processing algorithms have permitted 3-D images to be constructed (fig. 8.10).

NAVIGATION RADAR

Radar sets that assist navigation need not be as sophisticated as the ATC or terrain-mapping radars of figures 8.9 and 8.10. Much simpler, smaller, more economical radars can effectively aid navigation of aircraft or marine vessels.[17] Navigational radar sets usually operate over a relatively short range (to the horizon, for a surface vessel) and with limited signal processing yet provide valuable data. Radar works at night and in all weather conditions (the operating wavelengths of navigational radar units are chosen so that they penetrate through air and precipitation). Modern marine

17. Marine radars are usually S-band. That is, they operate over a short-wave band (frequencies of 2–4 GHz and wavelengths of 8–15 cm). Some marine radars work at even smaller wavelengths; the X-band (X for secret—the nomenclature dates back to World War II) operates at around a 10-GHz frequency and a 3-cm wavelength; these radars are short range but with improved bearing estimation accuracy.

FIGURE 8.10. A NASA X-band SAR radar image of Haiti, with 3-D rendering (vertical scale exaggerated by a factor of 2). This image was taken before the devastating earthquake of 2010; a linear fault line is evident in the center of the picture. Image courtesy of NASA/JPL/NGA.

radar equipment is smaller, more reliable, more capable, and easier to use than at any time in the past. Additional equipment enhances radar navigation. Thus, the humble *corner reflector* attached to a marker buoy—placed in a channel or harbor entrance, for example—reflects a signal back to a radar receiver. Corner reflectors work as shown in figure 8.11. The equipment is inexpensive, has no moving parts, and reflects a wide bandwidth of transmitted frequencies.

By now you have a pretty good idea of how radar can enhance marine navigation. Radar corner reflectors replace the lighthouses and other prominent coastal features used in earlier times; a ship offshore can obtain a position fix from two bearings to reflectors of known position. Better: because radar provides an estimate of target range, a single reflector (marked on a chart) will suffice to fix ship position via range and bearing. Another technique is *tangent bearing*. Suppose that there is no reflector within the operating range of your radar unit,[18] but you, cozily retired on your luxury

18. The effective operating range of a radar depends on many factors, such as transmitted power, transmitter and receiver beamwidths, target size, and, perhaps most important of all, the filtering and other processing that is applied to the received signal.

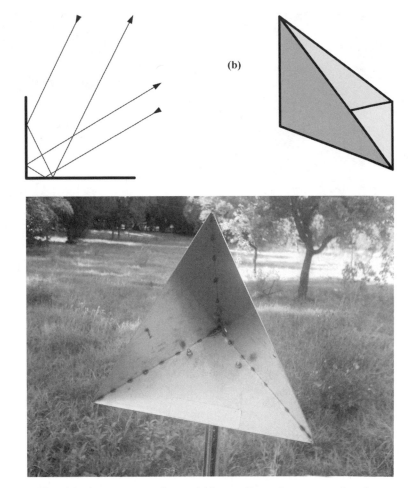

FIGURE 8.11. The ubiquitous corner reflector. (a) In two dimensions, we see how it works: whatever the direction of an incoming radar signal, the reflection is sent straight back to the radar. (b) In three dimensions, the corner reflector is simply a triangular corner cut from a hollow cube. (c) A trihedral corner reflector used for testing radar. Photo from Wikimedia Commons, User: Elborgo CC by 3.0.

yacht and cruising around the coast of Australia (whose hazards you may recall from chapter 7), observe a small island on your radar screen. You find the island on a map or chart and note from the radar data its general direction. The island may cover a wide range of angles, but the edges are well-defined. The direction and range of one edge are enough for you to fix your position on a chart.

Charts and maps can be used in conjunction with radars in another way. A map that is printed on a transparent sheet can be laid on top of a suitably scaled radar screen which displays, let us say, a group of islands. When the map and display image coincide, then the pilot or navigator can readily transfer his location (the center of the radar display) onto the map. This plot-overlaying positioning technique is known as *Fischer plotting*.

More sophisticated overlaying of different data goes by the fancy name (in military circles) of *sensor fusion*. Data from different electronic sensors can be downloaded directly to a radar display and placed upon it. Thus, for example, magnetic bearing from a fluxgate compass can be displayed on a screen, or data from LORAN or GPS can be overlaid. High-end radars may have a database of cartographic information from a GIS that is placed on a screen and updated as the coverage area being displayed on the screen changes. The database may show coastlines, for example, and these will confirm that certain of the radar signals on the screen correspond to land clutter, and that others correspond to known electronic beacons or reflectors.[19]

Radar leads the way in terms of electronic remote sensing and shipborne (and airborne) navigation equipment. Pilots depend on electronic navigation equipment as they never have before. They have learned to trust it because it has become so sophisticated and generally reliable, to the extent that it has rendered earlier mechanical or optical equipment redundant, except as a backup and for hobbyists. Yet, over the past decade we are seeing that radar itself is beginning to be overtaken by another form of electronic navigation—GPS. There are advantages of radar that will prevent its total eclipse, however. In certain circumstances navigation involves more than knowing your own position; for example, in a crowded harbor entrance in poor visibility, it is useful to also know the positions of everyone else.[20]

19. The use of radar for maritime piloting is well explained in chapter 13 of Bowditch's *The American Practical Navigator*, a long-running U.S. government marine navigation guide. See www.irbs.com/bowditch.

20. An anecdote that I have told elsewhere and often, which I first heard in 1990 (before GPS), involves an overreliance on newfangled electronic harbor navigation equipment. A Danish merchant marine captain told me (in a pub on the Orkney Islands off northern Scotland) of a sea trial of new equipment which, it was claimed, enabled a pilot to bring his boat into harbor without the need for any visual cues. A trial ensued in which the pilot was below deck, staring at a radar display—his only source of information about his external surroundings—and attempting to bring his boat to dock. In fact, he crashed the boat into a

GPS

We encountered GPS in chapter 3 and saw how the system worked. Here, we look at it from the other end—from the user's perspective—and then consider where it will lead us in future.

Hand-held GPS devices (fig. 8.12), or dashboard-mounted instruments with their route calculators and soothing voices, are commonplace and even passé. The latest word in personal positioning and navigation systems —but certainly not the last, as I will argue in this section—is the cell-phone app. For example, a free Google application, beta-released at the time of writing, provides Google Map pictures of your immediate location on your phone, instead of crude line drawings we see on a dedicated GPS receiver; this app guides you to your destination while performing Internet searches to provide the latest information, such as traffic jams and road-work updates.

Even more sophisticated is a new GPS navigation application for up-scale automobiles, called Virtual Cable, already field-tested by a San Francisco company. It projects a head-up display onto your windshield, much as a fighter pilot is presented with a HUD of vital information that allows him to keep his eyes on the skies. The difference is that Virtual Cable shows, not incoming missiles, but a 3-D image of a red cable suspended above the road you are driving along, receding into the distance in true perspective and turning left or right where you are to turn left or right. This form of route guidance is very intuitive and presents the navigation directions in a manner that is readily taken in by the driver. It's also undeniably very cool.[21]

The applications of GPS are spreading across our lives like ink on a blotting paper as we begin to absorb the full potential of this satellite technology. Of course, in the future we will continue to use GPS to determine our location, whether we are hiking in the hills far away from civilization or searching for a suburban restaurant. We will continue to seek directions while in our car. Ambulance drivers, taxi drivers, and express mail delivery companies will continue to use GPS to quicken their journeys. Air traffic controllers will depend upon GPS to maintain safe distances between airplanes at busy airports. Tomahawk missiles will make

---

small harbor lighthouse (an irony), which then collapsed onto the boat. The pilot then had to suffer the indignity of manually guiding his vessel into port with the wreckage of a lighthouse strewn across the deck.

21. Details of this instrument, and videos of a field trial, can be seen at the website www.mvs.net.

FIGURE 8.12. A TomTom 6 handheld GPS receiver. Wikipedia.

use of the military version of GPS (let us not forget the origin and motivation for this satellite system) to guide them to their primary targets,
redirect them mid-flight to a secondary target, keep them updated as to the
position of a moving target, or maintain the exact location of a target as
the missile loiters above it, perhaps for hours, waiting for that bunker
door to open.

In the future, the use of GPS by aircraft, automobiles, ships, airplanes,
and farm vehicles will increase. Farm vehicles? GPS can be used to identify
unmarked boundaries and problem areas in large fields of crops. Fishing
boat crews will use GPS to mark the locations of crab pots. The accuracy
and reliability of GPS will improve (despite U.S. Air Force warnings about
GPS funding and satellite maintenance issues). The NAVSTAR satellites

that are the backbone of GPS will be integrated with the GALILEO satellites of the equivalent European system. The next generation may see GPS replaced by GNSS (Global Navigation Satellite System). Businesses will increasingly employ smaller and smaller GPS receivers to track their vehicles and monitor their employees and as anti-theft devices. GPS receivers will be attached to wandering children in day-care centers and to confused retirees in hospitals. There will be privacy and civil liberties issues as we deplore the Big Brother applications of our employers, while checking online to see just where that pizza delivery guy is.

Let us not lose sight of the marvel of GPS, as the apex of thousands of years of striving to improve our navigational capabilities. For a few hundred dollars (or pounds, euros, yen, yuan, or rupees) we have at our fingertips a device that performs far, far better, and is much easier to use, than the high-tech navigational instruments of the past. Think of the vast fortunes that were offered for the inventor of an instrument that could measure longitude to within one nautical mile (1852 m). Now for a small fraction of the cost of John Harrison's H4 marine chronometer, we have a receiver that tells us both our latitude and longitude to within 1% of that distance.

Nature's Navigators

We have seen how the ancient craft of navigation has evolved over millennia from humble beginnings into the very high-tech, high-precision science that it is today. I might have chosen to summarize this evolution here in the closing section by simply listing our modern navigational capabilities, or perhaps by emphasizing how these capabilities have exploded exponentially over the last 20 years. This manner of wrapping up the book would have been reasonable, if a little uninspired. It seems to me, however, that there is a better wrap-up—one that conveys the breadth of navigational technology that we humans have built up over time and yet places these achievements in a strange and rather interesting perspective.

We humans are not alone in our need for navigation and our ability to navigate. Some animals have to navigate accurately over long distances, and their skills and capabilities have evolved over eons to a degree that is quite astonishing. Most of us are familiar with the barn swallow, which migrates from under our roofs to some far-away wintering ground in the southern hemisphere (I am writing as a resident of the northern hemisphere) and then returns to the very same nest site the following summer. Clearly, this accomplishment requires significant navigational skill. A more detailed investigation of animal navigation reveals that these birds are just the tip of a large metaphorical iceberg: other birds and animals exhibit, between them, *all* of the navigational capabilities that we humans have developed with our technology. Indeed, I will go so far as to say that there is nothing we humans have achieved in this field that is not matched somewhere in the animal world.

There is an eerie analogy between the systems that we use to aid navigation and the systems that have evolved in the natural world. (Maybe it's not so eerie; after all, everyone and everything is subjected to the same laws of physics, so perhaps we should expect that they will come up with similar solutions.) So I conclude my account of the history of human navigation by looking at examples of matching navigational techniques in the animal world. Odd, I know, but revealing and somewhat humbling.[1]

Piloting. Recall the Phoenicians and Carthaginians, boldly going where nobody had gone before (except for the residents of the African coasts that they skirted) in their wooden sailing ships. The pilot of such a ship could easily retrace his steps. That is to say, he could follow familiar landmarks to his home. Doing this requires a map (perhaps physical or perhaps merely a mental map) on which each landmark has its place relative to the home port and from which the pilot can guide his ship home.

There are 130 species of digger wasps in existence; the females of some species use piloting to find their way back to their nests. The nest entrances are not visible from far away, so a wasp memorizes the local topography and uses the mental map it has formed as a guide. We know this because, in an experiment that is well known to zoologists, researchers moved some prominent landmark features (such as pebbles) that were near one nest site while the wasps were elsewhere. The returning wasps had difficulty finding the nest entrance and took much more time to do so than they had needed before the landmarks were moved.

Compass. Piloting requires a map but not a compass. Humans have developed a magnetic compass to indicate direction and have learned how to navigate with this tool. So have birds. The main difference is that the birds' compass is built-in. In fact, there are many different compasses within the animal world. One popular type (in the sense that many species have evolved it) is based upon sensitivity to the geomagnetic field. Some birds have small particles of natural magnetic material in their brains that provide them with a sense of direction, defining their orientation with respect to the geomagnetic field. For example, consider the European cuckoo. As soon as it has learned to fly, a young cuckoo migrates from its nest in, say, western Scotland to wintering grounds in central Africa. It flies alone; given the nature of cuckoos, we can hardly expect its parents to show it the way. These young birds fly for thousands of miles over regions that they have not visited before, and yet they all end up in the same broad area that

---

1. For more on the fascinating subject of animal navigation, see Denny and McFadzean (2011). For animal remote sensing, see Denny (2007, chap. 5).

provides them with suitable habitat for the winter. Clearly, they are following an internal, hard-wired program that tells them, for example, "Fly south for 1,400 miles; turn east for 300 miles to avoid the desert; resume a southerly course until you run out of steam." Such an internal set of instructions works because the bird possesses a compass.

Dead reckoning. We have seen that during the Age of Exploration, European navigators crossed the world's oceans for the first time and that, until they solved the longitude problem, they estimated longitude by dead reckoning. Indeed, dead reckoning was used more generally for a ship out of sight of land to keep track of its position. There are insects that have been observed to carry out very similar piloting tasks. Sahara Desert ants live in a world that is quite featureless. They leave their nest burrows in a wide search for food; when they find some, they take it back to their nests. The impressive aspect of their movement is that when it is time to return home, they immediately head in the right direction and make a beeline for the nest, even if it is out of sight. They are applying dead reckoning to keep tabs on the distance and direction home.

Dead reckoning requires a compass and a clock, to estimate speed. Biological clocks are very common—you and I have them—but the desert ant also has a sun compass. The sun can be relied on to shine in the Sahara Desert. With an internal clock to compensate for time of day, the ants can use the sun to estimate direction. They also use an internal clock to estimate the distance they travel for a given speed. When they change direction, they record the new direction and distance traveled along it. They add vectorially all the directions and distances (this is the path integration of fig. 6.12a), so that when it is time to go home, they know the way.

Celestial navigation. We have taken advantage of a Pole Star for many millennia and have learned to be guided by it. Today, Polaris is the closest star to the earth's North Pole, meaning that we see it rotate with the smallest circle as we watch the night sky circle above us. Nestling birds do the same thing, or at least they do so if they are nocturnal migrants. Nestlings need to learn the night sky, just as mariners in medieval Europe or Polynesia learned the night sky, and for the same reason: Polaris tells them where north is, and the visible constellations and where they rise and set tell these birds their own latitude. We know this because of experiments with birds in planetariums. In a planetarium, the night sky can be changed so that, for example, Betelgeuse and not Polaris is nearest the rotational center. Nestlings who are raised under such a sky, when released for their first annual migration, head in a direction that is based upon the assumption that Betelgeuse is due north.

Furthermore, there is a phenomenon that has long been observed in nocturnal migrant birds that have been confined to cages. It is termed *migratory restlessness*, and it manifests itself as follows. A caged bird that should be migrating at a certain time of year is seen to flutter toward the migration direction—that is, it spends most of its time in the part of its cage that is in the direction of its natural migration route. If a migrating bird is known to change direction after, say, 10 nights of flying, then a caged migrant will also switch its flutter direction after 10 nights. Migration restlessness stops after a period of time that is the same as the duration of the migrant's journey. Clearly, what is going on here is the acting out of an internal program.

The point is that, in the context of navigation, the birds are demonstrating several features required for celestial navigation. They learn the stars and steer by them. If the stars are artificially changed, then they learn in the wrong way, but it is exactly the same wrong way as a human navigator would learn under the same artificial night sky. They have an internal clock, so they know about nightly changes in star patterns during the migration season. They know when to stop migrating either by consulting an internal clock or because they have spotted a constellation that is in the right part of the sky, at the right time, at the latitude of their winter quarters. Amazing, for a birdbrain.

Radar. OK, so no animal apart from *Homo sapiens* uses radar. However, radar and sonar are very similar, and in fact, they rely upon the same physical principles and use many of the same processing algorithms. Within the animal world there are two groups that have developed sonar to a degree of sophistication that matches or exceeds our own: the Microchiropteran bats (a large group with over 800 species that includes most of the small bats that fly above your house at night hunting insects) and the toothed whales (such as dolphins). Engineers have tuned up our radar accuracy and precision using various mathematical techniques such as correlation processing and with broadband methods such as chirp processing. Bats and toothed whales do exactly the same.

Experiments show that both these types of creature are able to create a *sound map*—an acoustic image in their brains—in the same way as our eyes present data to our brains that we process into a visual image of the world around us. We have learned to navigate by creating radar maps based upon the movement of the radar platform—SAR images. The acoustic equivalent used by bats must be based upon different principles that we don't yet understand because they can make maps from what they detect directly in front of them. (Our SAR maps work only when a moving radar antenna looks out to the side; the map degrades in forward directions to the extent that it is useless when looking straight ahead.) All the techniques that we use in our radar and sonar signal processing (such as beam-forming, Doppler processing, and signal integration) are also used by bats and toothed whales. Researchers have recorded the details of sonar transmissions from these animals and, making use of their considerable knowledge of signal processing, have calculated the theoretical best resolution and accuracy that can be attained. A bat uses its sonar to locate a flying insect in the dark; a dolphin uses it to locate a fish in the sea. The accuracy and precision that bats and dolphins achieve is the best possible—they are as expert as we are.

GPS. Surely animals do not have satellites and GPS! Satellites, no—but GPS, yes. Again, birds are the impressive exemplars. Their GPS system is provided by the earth itself (in some cases the sun also plays a role); and a bird that travels long distances, either on migration or to gather food, uses it with consummate skill to determine its latitude and longitude. We know this from a series of displacement experiments in which birds have been captured, transported in a covered cage to a destination many miles distant, and then released. These birds find their way home very quickly.

A famous example happened in the 1950s, when some Manx shearwaters (pelagic birds that nest in coastal burrows) were removed from their nest site along the west coast of Wales and transported by air across the Atlantic to the Massachusetts coast. They were released in a part of the world they did not know—the Eastern Seaboard is not part of their regular stamping ground—and yet they found their way home in less than 13 days. Indeed, the first arrival got home before the letter announcing its release reached Wales from Massachusetts.

There are many other examples in which researchers drugged or blindfolded birds during transit to ensure that they were not making use of cues and doing some sort of sophisticated dead reckoning. It is certain that these birds, when released, were in parts of the world they had never seen before (for instance, an albatross from the southern hemisphere was taken to Whidbey Island in the Pacific Northwest), and yet they unerringly returned home to their nest sites.

This level of skill requires a map and a compass and more: a map is useless if you do not know your position on it. Displaced birds call upon their built-in GPS system to tell them their latitude and longitude. Latitude is easy to gauge from the geomagnetic field because, recall, the dip angle varies with latitude. Birds can estimate longitude in a number of ways, such as by comparing their internal clock with local sunrise, or by listening to the earth. Huh? Some birds are sensitive to infrasound—very-low-frequency sound waves, far below the lowest frequencies that we humans can perceive. Naturally produced infrasound arises from *microseisms*, the rumblings of our oceans. Infrasound waves travel thousands of miles, and microseisms travel inland long distances. To a bird that can hear infrasound, the world is full of acoustic landmarks. The shape of continents tunes the infrasound, making it different in different places, so that birds may know their location on the earth just from the infrasound they hear.

Sensor fusion. Migrating birds in particular are good at navigating because they use many different techniques. Just as we might use GPS and back it up with a radar reading, a celestial observation, or a sextant measurement, so birds may use their magnetic sense and back it up with celestial observations, infrasound, or a sun compass. When I worked in the avionics industry, we referred to the pooling together of navigational data from different sources as *sensor fusion*.

The European robin exhibits a particularly nice example of sensor fusion. It probably can *see* the geomagnetic field, perhaps as a turquoise shading of the sky getting deeper toward the direction of the North Pole. Here, optical and geomagnetic information is being blended; in particular, the geomagnetic data is being displayed in the mind's eye of the bird (through its right eye only, researcher have discovered). This display constitutes a very sophisticated system that presents complex data in a manner that can be meaningfully grasped quickly.

There are many other examples of sensor fusion in navigating birds. For example, pigeons and many other species are sensitive to polarized light and to the geomagnetic field. Polarized light enables them to sense the direction of the sun even after it has set or while under clouds. The birds may prefer the geomagnetic sense in some circumstances and the polarized light information in others. For example, if a bird suspects its magnetic reading are false (maybe because of a local magnetic anomaly or because an ornithologist has placed the bird in an artificial magnetic field), it will recalibrate the direction it senses magnetically, using the direction of north as indicated by polarized light. Or it may switch from celestial navigation, or from its sun compass, if clouds block the view, to magnetic sensing or infrasound.

Many animals exhibit versions of all the various techniques for navigation that we have developed through mathematics and technology over the past 3,000 years. Both humans and animals have watched the stars to know where north is; we have both developed clocks and sun compasses, geomagnetic compasses, and sonar (and in our case also its electromagnetic equivalent, radar). We have both developed mapping skills and piloting, map-and-compass skills and dead reckoning algorithms, and have learned to pool together navigational data from different sources. The technological skills of humans are directed—they are learned and improved upon by intelligent application—whereas animal navigation evolved slowly over evolutionary time, without direction. Thus, humans are poised to take over the lead in this field. I wonder what the next 3,000 years will bring.

Technical Appendix

The Mercator Projection and the Gyrocompass

In this appendix I break my rule about presenting nonmathematical elucidations of navigational science in this book. I made this rule, you may recall, because the math involved in our subject can get a little hairy and, frankly, quite tedious. However, the Mercator projection is historically important, and explanations of how the gyrocompass works are often not accurate.

The Mercator Projection

By showing you the mathematical derivation of the Mercator projection, I am conveying the essence of cartography: projecting points on a sphere onto points on a plane. Each of the projections we saw in chapter 4 has a different formula, some of which are much more complicated than the Mercator. Here you have the full derivation.

Consider figure A1. The Mercator projection is a cylindrical projection, so that the earth is mapped onto a cylinder that can then be unrolled to form a flat rectangular map. In figure A1(a) we have wrapped the earth (assumed to be spherical) with a large piece of paper, which forms a cylinder touching the globe at the equator. The essence of the Mercator projection is easy to describe. First, each point on the equator of the globe is already mapped onto a point on the cylinder, defining the equator on our map. Now let us project point A on the globe onto point A' on the cylinder. The large arrow in figure A1(a) shows the projection of A to A'. Imagine a plane that passes through three points on the globe: the North and South Poles and the point A. This plane will cut the cylinder along a straight line. The line will be vertical and, when the cylinder is unfolded, will be a meridian of the map. In this way, points on the globe are easily projected onto the correct longitude on the cylinder.

What about latitude? The plane cuts the globe as shown in figure A1(a). The length of the cut from the equator to the point A is s. On the cylinder, we measure from the equator northward (in this case, because A is north of the equator) a distance s and mark the point A' on the cylinder. In other words, the distance on

(a)   **(b)**

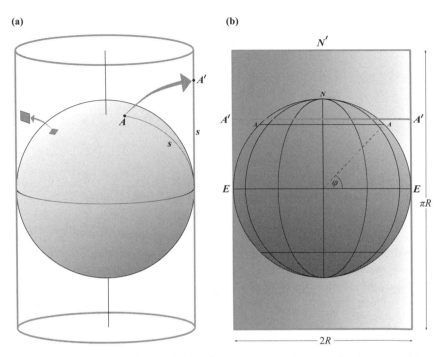

FIGURE A1. Mercator projection. (a) A cylinder is wrapped around the globe, touching it only at the equator. Point *A* on the globe is mapped onto point *A'* on the cylinder. The latitude of *A'* is determined by the distance, *s*, of *A* from the equator. A small area on the globe maps onto a small area on the cylinder. (b) The cylinder dimensions are shown; it is just long enough to map all latitudes. The North and South Poles map onto the top and bottom edges of the cylinder. The distance *A'E* (let us call it *y*) on the Mercator map is determined as a function of the latitude φ of *A*: see equation (A1).

the cylinder of point *A'* from the equator is just the arc length on the globe of point *A* from the equator. This is the Mercator projection, and it is easier to picture than to explain in words.

Another way of viewing the Mercator projection is to consider the globe to be a balloon. Inflate the balloon so that it is squeezed up against the side of the cylinder: *A'* is the point on the cylinder that is adjacent to the point *A* on the inflated globe.[1]

In figure A1(b) you can see the same cylinder, wrapped around the earth, viewed from the side. The height of the cylinder—and so the height of the map—is πR, where *R* is the earth's radius. The width of the map, when the cylinder is unrolled so that the map is flat, is twice the height, $2\pi R$, which is, of course, just

1. An animation of this inflated-balloon view of Mercator projections can be seen at the University of British Columbia's website at www.math.ubc.ca/~israel/m103/mercator/merca tor.html, where a math derivation is also provided.

the earth's circumference at the equator. Now scale the map down—say, by a factor of 6,336,000—and we have a Mercator projection map of a sensible size (in this case, 1 inch represents 100 miles). Note from figure A1(b) one disturbing feature of the Mercator projection: the North Pole is mapped onto a line (the top edge of the map), not onto a single point. The South Pole is similarly mapped onto the bottom edge of the map. This reflects the distortion of Mercator maps at high latitudes.

So much for the geometry: now for an algebraic analysis of Mercator projections that provides us with the mapping formulas and also shows that this projection preserves angles. We want to project a location on the globe, say at latitude ϕ and longitude λ, onto a position (x,y) on the map. Mapping longitude is easy, as we have seen: mathematically, we say that $x = \lambda$, meaning that the horizontal coordinate on the map is just the angle of longitude. Mapping latitude is trickier. Figure A1(a) shows a small square on the globe that is mapped onto the cylinder. To preserve angles, the square must be mapped to a square so that the local shape of things is maintained. Say the center of the square is at latitude ϕ on the globe; you can see from figure A1(b) that if the globe is of radius R, then the parallel at latitude ϕ is a circle of radius $R \cos \phi$. On the map, parallels are horizontal lines of equal length (and meridians are vertical lines of equal length) because the map is rectangular. Thus, parallels at latitude ϕ on the globe must be stretched by a factor of $1/\cos \phi$ when they appear on the map.

Let us say that the small square of figure A1(a) has sides that are along meridians and parallels. If the square extends over a small angular range of longitudes, say $d\phi$, then the length of each side of the square is $R\, d\phi$. But the parallels at latitude ϕ are stretched, and so, on the map, the square is of width $R\, d\phi/\cos \phi$. (This stretching must occur if the square is to remain a square when mapped.) I will denote this small width dy. Thus, $dy/d\phi = R/\cos \phi$. This equation is a differential equation for the vertical coordinate (y) of a point on the map. It can be integrated (solved) to yield $y = R \ln[\tan (\frac{1}{2}\phi + \frac{\pi}{4})]$. Thus, the formula for a Mercator projection is as follows:[2]

$$(x,y) = (\lambda, \ln[\tan(\tfrac{1}{2}\phi + \tfrac{\pi}{4})]). \tag{A1}$$

Note that the analysis presented here has made use of logarithms and of calculus. Neither of these mathematical ideas was known when Edward Wright provided the first mathematical analysis of the Mercator projection in 1599. The following century brought more elegant analyses, by James Gregory and Isaac Barrow.

In figure A2 you can see what a rhumb line looks like on a linear scale (no projection) and on a Mercator projection map. Here, I assume that a course is followed that cuts meridians at 45°. The bearing must change on the linear

---

2. The formula for an oblique Mercator projection is more complicated. See, for example, Eric W. Weisstein, "Mercator Projection," at the website MathWorld—A Wolfram Web Resource, at http://mathworld.wolfram.com/MercatorProjection.html.

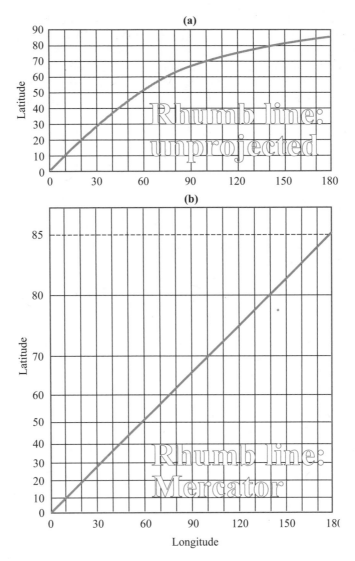

FIGURE A2. The difference between a rhumb line on an unprojected map (a) and on a Mercator map (b), assuming that the course direction is 45° north.

map (to account for converging meridians), but on the Mercator map, the line is straight.

The Gyrocompass

A gyrocompass is shown conceptually in figure A3. The axis, *CD*, of a spinning disk is permitted to move freely, but another axis, *AB*, is constrained to remain in the

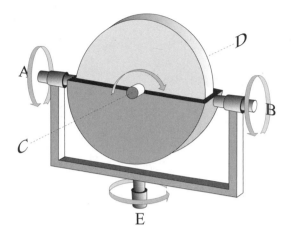

FIGURE A3. Conceptual model of a gyrocompass. The axis *AB* is constrained to lie in a horizontal plane because the axis *E* is vertical—that is, it is pointing to the center of the earth. The gyroscope spin axis is *CD*. A heavy case weights the gyroscope so that if no torque acts, the spin axis is horizontal.

horizontal plane. Imagine that the axis *E* of this figure is fixed to a pole that is placed vertically in the ground. The lower part of the spinning disk is enclosed in (but does not touch) a heavy case. The case serves to orient the disk so that, under the action of gravity, the spin axis, *CD*, will be horizontal. It is crucial to note, however, that the disk is free to rotate about the axis *AB* if a torque is applied.

To see how a gyrocompass works, we can employ this conceptual model as shown in figure A4.[3] Imagine our conceptual gyrocompass on a pole that is stuck vertically in the ground, initially at position *a*. The heavy case, which is pulled by gravity, ensures that the spin axis, *AB* of figure A3, is horizontal. A few hours later, the earth has rotated the gyrocompass to position *b*. However, because of its propensity to stay in the same orientation (due to the large angular momentum of the fast-spinning disk), the spin axis is still horizontal. Gravity acts on the case to pull it down, thus exerting a torque about the axis *AB* of figure A3. The direction of the torque is out of the page, toward you.[4] Torque acts to change angular momentum, so the direction of the spin (dashed arrows in fig. A4) is changed: the arrow moves north, out of the page. Here is the gyrocompass beginning to do its thing—aligning itself with the earth's spin axis by pointing north.

3. Real gyrocompasses do not resemble the model of figure A3, which is only for pedagogical purposes. The explanation of gyrocompass behavior given here is close to that of a very early U.S. Navy report on this device; see Gillmor (1912). For more information on modern gyro technology, see Lawrence (1998). There is also a helpful online account of gyroscope use in maritime navigation on a page of the San Francisco Maritime National Historic Park website, at www.maritime.org/fleetsub/elect/chap17.htm.

4. This behavior is surprising: we expect the torque to pull the gyroscope as shown by the curved arrow at *b* in fig. A4. Instead, it acts to pull the gyroscope out of the page. This odd behavior is understood by physicists but is counterintuitive. The same behavior makes a child's spinning top precess instead of fall over. The explanation lies in the manner in which angular momentum vectors add: the *direction* of the torque is out of the page, according to the right-hand rule discussed in the box "Spinning Tops and Angular Momentum" in chapter 8. This torque vector is added to the gyroscope spin vector to produce the change of direction shown in fig. A4.

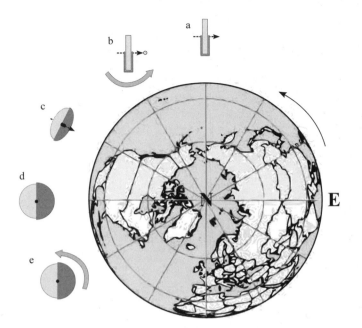

FIGURE A4. A conceptual gyrocompass is shown edge-on at its initial position *a*. The gyrocompass reorients itself during the day, as the earth rotates, so that its spin axis points north-south. (At *b* the small circle with a dot in the center represents an arrow pointing toward you; it is a component of gyroscope angular momentum that is generated by gravitational torque.) The torque that causes this reorientation is weakest at polar latitudes and strongest at the equator.

The gyrocompass now finds itself at position *c* and then *d*, with its spin axis (*CD* of fig. A3) now pointing due north. Further earth rotation moves the gyrocompass to position *e*. Now you can see that the action of gravity exerts a torque that is directed out of the page, as before, but this does not change the axis orientation, which is already pointing in that direction. Thus, gravitational torque acts to orient the axis so that it is parallel to the earth's rotation axis, but once the axis is so oriented, the torque has no further effect, and the gyrocompass remains oriented north-south.

It is the torque due to the force of gravity acting on the heavy case that causes this reorientation of the gyroscope, turning it into a compass. Without the heavy case there is no torque, and the gyroscope axis would remain in the orientation of its initial position, *a*, all day.

The gyrocompass points to true north, not magnetic north, but it is not perfect. The torque that brings about the axis reorientation is weak at polar latitudes because the component of gravitational force acting in the plane of the paper in figure A4 is weak at these latitudes. In practice, this means that gyrocompasses do not work well at very high latitudes, north or south. Also, a gyrocompass of this design will not work well in an airplane that is buffeted around (by stormy weather, for example). It will also not work in an airplane that is flying east at a high speed that corresponds to the speed of the rotating earth because then the gyrocompass will remain at position a as the earth rotates beneath it and will feel no torque. This instrument was originally designed for use on board big ships, which move slowly. Modern versions of the gyrocompass, however, do work well on airplanes.

Annotated Bibliography

Well, partly annotated: I comment on those references that I found to be particularly helpful, those that are a good read, and also those that are broad enough in scope to touch upon more than one of the subjects discussed in this book. As I remarked in the introduction, I am hoping that the main text will have whetted your appetite for further reading on some of these subjects; the annotated texts will take you to a deeper level.

Acar, C., and A. Shkel. 2008. *MEMS Vibratory Gyroscopes*. New York: Springer.

Adam, D. 2002. "Gravity Measurement: Amazing Grace." *Nature* 416:10–11.
   A non-technical overview of a very technical subject.

Allen, O. E. 1980. *The Pacific Navigators*. Alexandria, VA: Time-Life Books.
   A very readable account of the European exploration of the Pacific from the late sixteenth to the late eighteenth centuries.

Ambrose, S. E. 2003. *Undaunted Courage*. London: Simon & Schuster.
   An engaging if somewhat rhapsodic account of the Lewis and Clark expedition from St. Louis to the Pacific coast, in the first decade of the nineteenth century. This account sets the table first, with useful information on the political background to the expedition.

Anderson, E. W. 1997. "The Treatment of Navigational Errors." *Journal of Navigation* 50:362–71.

Andrewes, W. J. H. 1993. *The Quest for Longitude*. Cambridge, MA: Harvard University Press.
   A comprehensive, balanced, and engaging account of the long search for a method of estimating longitude.

Arnold, D. 2002. *The Age of Discovery, 1400–1600*. 2nd ed. London: Routledge.

Avery, T. 2009. *To the End of the Earth*. New York: St. Martin's Press.

Balchin, J. 2004. *Quantum Leaps: 100 Scientists Who Have Changed the World*. Wigston, Leicester, England: Arcturus Publishing.
   Biographical sketches, at least ten of which are relevant to our story.

Bartlett, T. 2009. *The Book of Navigation: Traditional Navigation Techniques for Boating and Yachting*. New York: Skyhorse Publishing.

Bawlf, S. 2003. *The Secret Voyage of Sir Francis Drake*. Vancouver, BC: Douglas & McIntyre.

An interesting and controversial version of the epic journey of Francis Drake around the world. Written by a former politician, it posits political intrigue and reads like a thriller.

Bedini, S. A., ed. 1998. *Christopher Columbus and the Age of Exploration*. New York: Da Capo Press.

A really excellent encyclopedia covering every conceivable topic relating to the European Age of Exploration from the late fifteenth to the mid-seventeenth centuries.

Beer, A., et al. 1961. "An 8th-Century Meridian Line: I-Hsing's Chain of Gnomons and the Prehistory of the Metric System." *Vistas in Astronomy* 4:3–28.

Berggren, J. L., and A. Jones. 2000. *Ptolemy's Geography: An Annotated Translation of the Theoretical Chapters*. Princeton, NJ: Princeton University Press.

Bergreen, L. 2007. *Marco Polo—From Venice to Xanadu*. New York: Alfred A. Knopf.

A detailed account of the historically influential wanderings of the famous Venetian merchant.

Boorstin, D. J. 1983. *The Discoverers*. New York: Random House.

Bouvier, A., and M. Wadhwa. 2010. "The Age of the Solar System Redefined by the Oldest Pb-Pb Age of a Meteoritic Inclusion." *Nature Geoscience* 3:637–41.

Breitenberger, E. 1984. "Gauss's Geodesy and the Axiom of Parallels." *Archive for History of Exact Sciences* 31:273–89.

Brown, D., ed. 2003. *The Greatest Exploration Stories Ever Told*. Guildford, CT: Lyons Press.

Buffet, B. A. 2000. "The Earth's Core and the Geodynamo." *Science* 288:2007–12.

Burnham, R. 1979. *Burnham's Celestial Handbook*. New York: Dover.

Buttimer, A., and L. Wallin, eds. 1999. *Nature and Identity in Cross-Cultural Perspective*. Dordrecht, The Netherlands: Kluwer Academic.

Campbell, T. 1981. *Early Maps*. New York: Abbeville Press.

Carrigan C. R., and D. Gubbins. 1979. "The Source of the Earth's Magnetic Field." *Scientific American* 240 (Feb.): 118–29.

Chambert, J.-L., ed. 1999. Chapter 9 in *A History of Algorithms*. Berlin: Springer-Verlag.

Chevallier, R. 1984. "The Greco-Roman Conception of the North from Pytheas to Tacitus." *Arctic* 37:341–46.

An interesting academic treatise that analyzes the influence of Pytheas in shaping the world view of classical antiquity.

Clausing, D. J. 2007. *The Aviator's Guide to Navigation*. 4th ed. New York: McGraw-Hill.

Cook, A. 1998. *Edmond Halley*. Oxford: Oxford University Press.

The definitive biography, describing Halley's accomplishments and the times he lived in. Best known for his comet, Halley also investigated geomagnetism and tides.

Crane, N. 2002. *Mercator: The Man Who Mapped the Planet*. New York: Henry Holt.

A detailed biography of the man and a history of cartography.

Cunliffe, B. 2002. *The Extraordinary Voyage of Pytheas the Greek: The Man Who Discovered Britain*. New York: Walker & Co.

Cutter, T. J. 2004. *Dutton's Nautical Navigation*. 15th ed. Annapolis, MD: Naval Institute Press.

Daniels, J. E., and J. Wishart. 1951. "The Theory of Position Finding." *Journal of the Royal Statistical Society* 13:186–207.

Danson, E. 2006 *Weighing the World*. Oxford: Oxford University Press.

A very readable account of eighteenth-century developments in surveying and geodesy, and of the surveyors, mostly in England and France.

Davis, F. E., et al. 1998. *Surveying: Theory and Practice*. 7th ed. New York: McGraw-Hill.

An enduring standard textbook, first published in 1928.

Denny, M. 2007. *Blip, Ping, and Buzz: Making Sense of Radar and Sonar*. Baltimore: Johns Hopkins University Press.

Chapter 1 discusses electronic warfare in the Battle of Britain, including the Lorenz RDF system. Chapter 5 includes an account of animals' sonar signal processing techniques. The aim of this book is to show nonspecialists how radar and sonar work.

———. 2009. *Float Your Boat! The Evolution and Science of Sailing*. Baltimore: Johns Hopkins University Press.

My account of the evolving technology of sailing ships.

Denny, M., and A. McFadzean. 2011. *Engineering Animals*. Cambridge, MA: Harvard University Press.

Animal navigation is the subject of chapter 11.

Dent, B. D. 1998. *Cartography with ArcView GIS Software and Map Projection Poster*. New York: McGraw-Hill.

Duggal, S. K. 2004. Chapters 2–3 in *Surveying*. Vol. 2. New York: McGraw-Hill.

Dunn, R. E. 1989. *The Adventures of Ibn Battuta*. Berkeley and Los Angeles: University of California Press.

This fascinating account tells of the lengthy journeys across most of the fourteenth-century Islamic world by an inveterate traveler with a keen eye. The story tells itself: the author is happy to remain in the background.

Edmondson, A. C. 2007. Chapter 16 in *A Fuller Explanation: The Synergetic Geometry of R. Buckminster Fuller*. Pueblo, CO: Emergent World Press.

Edson, E. 2007. *The World Map, 1300–1492*. Baltimore: Johns Hopkins University Press.

Eley, T., and C. Northon. 2001. "Undercover Geography." *Geographical Review* 91:388–98.

El-Rabbany, A. 2002. *Introduction to GPS: The Global Positioning System.* 2nd ed. Norwood MA: Artech House.

A clear, nonmathematical description.

Encarta Encyclopedia. 2005. Standard Edition.

Encyclopaedia Britannica. 1998. CD 98 Standard Edition.

Evans, J. 1998. *The History and Practice of Ancient Astronomy.* Oxford: Oxford University Press.

Fernández-Armesto, F. 2006. *Pathfinders: A Global History of Exploration.* New York: W. W. Norton.

Fischer, I. 1975. "Another Look at Eratosthenes' and Posidonius' Determinations of the Earth's Circumference." *Journal of the Royal Astronomical Society* 16:152–67.

Fleming, F. 1998. *Barrow's Boys.* New York: Atlantic Monthly Press.

The engrossing story of nineteenth-century Arctic and African exploration by the men of the British Admiralty.

Garner, R. L. 1988. "Long-Term Silver Mining Trends in Spanish America: A Comparative Analysis of Peru and Mexico." *American Historical Review* 93:898–935.

Gillmor, R. E. 1912. "Theory and Operation of the Gyroscope and the Sperry Gyroscopic Compass." *United States Navy Institute Proceedings* 38:519–49.

Glick, T., S. J. Livesey, and F. Wallis, eds. 2005. *Medieval Science, Technology, and Medicine.* Oxford: Routledge.

Gonzalez, J., and T. E. Sherer. 2004. Chapter 4 in *The Complete Idiot's Guide to Geography.* Indianapolis: Alpha Books.

Gosch, S. S., and P. N. Stearns. 2008. Chapter 4 in *Pre-Modern Travel in World History.* New York: Routledge.

Haase, W., and M. Reinhold. 1993. *The Classical Tradition and the Americas.* Berlin: de Gruyter.

Hakes, H. 2009. *John and Sebastian Cabot: A Four Hundredth Anniversary Memorial of the Discovery of America.* Whitefish, MT: Kessinger Publishing.

Harley, J. B. 1989."Deconstructing the Map." *Cartographica* 26:1–20.

Harley, J. B., and D. Woodward. 1987. *The History of Cartography.* Chicago: University of Chicago Press.

Harris, S. J., and B. L. Grigsby. eds. 2008. *Misconceptions about the Middle Ages.* New York: Routledge.

Debunking much of the nonsense that we were brought up to believe, this well-researched and unusual book is a mine of useful information for history buffs.

Hayes, D. 1999. *Historical Atlas of the Pacific Northwest: Maps of Exploration and Discovery; British Columbia, Washington, Oregon, Alaska, Yukon.* Vancouver, BC: Sasquatch Books.

Great maps, each with interesting accompanying text.

Heath, T. L. 1981. *Aristarchus of Samos: The Ancient Copernicus.* New York: Dover.

Herman, A. 2005. *To Rule the Waves.* New York: Harper.

An American's-eye view of the rise and influence of Britain's Royal Navy, with interesting accounts of maritime navigation and of Francis Drake, James Cook, and William Bligh.

Hiraiwa, T. 1967. "On the 95 Per Cent Probability Circle of a Vessel's Position." *Journal of Navigation* 20:258–70.

Hoare, M. R. 2005. *The Quest for the True Figure of the Earth.* Aldershot, UK: Ashgate Publishing.

One to get out of the library: an expensive but comprehensive history of the last four centuries of geodesy.

Holmes, R. 2010. *The Age of Wonder.* New York: Vintage Books.

Hopkirk, P. 1995. *Trespassers on the Roof of the World: The Secret Exploration of Tibet.* New York: Kodansha International.

A short but enlightening and entertaining description told by an experienced writer who is knowledgeable about this part of the world.

Howarth, R. J. 2007. "Gravity Surveying in Early Geophysics, I: From Time-Keeping to Figure of the Earth." *Earth Sciences History* 26:201–28.

Howse, D. 1980. *Greenwich Time and the Discovery of the Longitude.* Oxford: Oxford University Press.

Hurley, F., et al. 2001. *South with Endurance.* London: Book Creation Services.

A coffee-table book of Shackleton's Antarctic expedition, with the photographs of Frank Hurley.

Jardine, L. 1996. *Wordly Goods: A New History of the Renaissance.* New York: W. W. Norton.

———. 1999. *Ingenious Pursuits.* London: Little, Brown.

A compelling account of the beginnings of the Scientific Enlightenment in Western Europe. Includes the French expeditions to determine meridian arc length.

Johnson, A. 2008. *Solving Stonehenge: A New Key to an Ancient Enigma.* London: Thames & Hudson.

A tale of archaeological research, well told by one of the archaeologists involved.

Kaler, J. 1996. *The Ever-Changing Sky: A Guide to the Celestial Sphere.* Cambridge: Cambridge University Press.

A comprehensive nonmathematical survey of celestial phenomena and their earthly effects.

Karl, J. 2004. *Celestial Navigation in the GPS Age.* Arcata, CA: Paradise Cay Publications.

Kavanagh, B. F. 2008. *Surveying: Principles and Applications.* Upper Saddle River, NJ: Pearson/Prentice Hall.

A good review of standard surveying methods.

Kearey, P., K. A. Klepeis, and F. J. Vine. 2009. *Global Tectonics.* Chichester, UK: John Wiley and Sons.

Keay, J. 2000. *The Great Arc.* New York: HarperCollins.

Largely a biography of Lambton and Everest, this account of the Great Trigonometric Survey is a good read.

Kelsey, H. 2000. *Sir Francis Drake: The Queen's Pirate*. New Haven, CT: Yale University Press.

Kemp, P., ed. 1976. *Oxford Companion to Ships and the Sea*. Oxford: Oxford University Press.

Useful for entries on marine navigational instruments, past and present.

Kibble, T. W. B., and F. H. Berkshire. 1985. *Classical Mechanics*. Harlow, Essex, UK: Longman.

King, A. D. 1998. "Inertial Navigation: Forty Years of Evolution." *GEC Review* 13:140–49.

Knights, E. M. 2001. "Navigation before Netscape." *History Magazine*, Oct./Nov.

Konvitz, J. 1987. *Cartography in France: 1660 to 1848*. Chicago: University of Chicago Press.

Lacroix, W. F. G. 1998. *Africa in Antiquity: A Linguistic and Toponymic Analysis of Ptolemy's Map of Africa, Together with a Discussion of Ophir, Punt and Hanno's Voyage*. Saarbrücken, Germany: Verlag für Entwicklungspolitik.

A very detailed technical analysis of the geography of Africa, as viewed from classical antiquity. Appendix 4 provides accounts of the Carthaginian explorations.

Lambeck, K. 1980. *The Earth's Variable Rotation: Geophysical Causes and Consequences*. Cambridge: Cambridge University Press.

Landes, D. S. 2000. *Revolution in Time: Clocks and the Making of the Modern World*. Cambridge, MA: Belknap Press.

Lawrence, A. 1998. *Modern Inertial Technology: Navigation, Guidance, and Control*. New York: Springer-Verlag.

Leick, A. 2004. *GPS Satellite Surveying*. Hoboken, NJ: John Wiley & Sons.

Letham, A. 2008. *GPS Made Easy: Using Global Positioning Systems in the Outdoors*. Surrey, BC: Rocky Mountain Books.

Lewis, D. 1994. *We, the Navigators*. Honolulu: University of Hawaii Press.

A thorough technical account of Polynesian navigation methods written by a trained navigator.

Lewis, M. J. T. 2001. *Surveying Instruments of Greece and Rome*. Cambridge: Cambridge University Press.

Includes a discussion of ancient tunnels and Roman roads, both of which required accurate surveying as a prerequisite—especially tunnels that were dug from both ends and needing to meet up in the middle.

Leuliette, E. W., R. S. Nerem, and G. L. Russell. 2002. "Detecting Time Variations in Gravity Associated with Climate Change." *Journal of Geophysical Research* 107:2112.

Lindberg, D. C., ed. 1978. *Science in the Middle Ages*. Chicago: University of Chicago Press.

Liu, J. 1997. *Survey Review for the Civil Engineer*. 2nd ed. Oxford: Oxford University Press.

Love, R. S. 2006. *Maritime Exploration in the Age of Discovery, 1415–1800*. Westport, CT: Greenwood.

Maloney, E. S. 2006. Chapter 17 in *Chapman Piloting and Seamanship*. 65th ed. New York: Hearst Books.

Mason, S. F. 1962. *A History of the Sciences*. New York: Macmillan.
A good summary of the development of the natural sciences. Chapter 21 covers the application of science to navigation.

McCarthy, D. D. 1991. "Astronomical Time." *Proceedings of the IEEE* 79:915–20.

McCarthy, D. D., and P. K. Seidelmann. 2009. *Time: From Earth Rotation to Atomic Physics*. Weinheim, Germany: Wiley-VCH.

Menzies, G. 2002. *1421: The Year China Discovered the World*. London: Bantam Press.
Controversial and probably wrong, but nevertheless interesting, this book tries to persuade us that the Chinese circumnavigated the world a century before Magellan. The book argues well that the Chinese had the nautical capability, but it fails to convince me that they chose to employ it as claimed.

Mills, G. B. 1980. "Analysis of Random Errors in Horizontal Sextant Angles." Master's thesis, Navy Postgraduate School, Monterey, CA.

Milton, G. 1999. *Nathaniel's Nutmeg*. New York: Farrar, Straus & Giroux.
A very readable history about the seventeenth-century conflict between the English and the Dutch over the Spice Islands.

Moffatt, K. 1993. "Cosmic Dynamos: From Alpha to Omega." *Physics World* 6 (May): 38–42.

Morrison, J. E. 2007. *The Astrolabe*. Rehoboth Beach, DE: Janus.

Mörzer Bruyns, W. F. J. 1994. *The Cross-Staff: History and Development of a Navigational Instrument*. Amsterdam: Nederlandsch Historisch Scheepvaart Museum.

Murdin, P. 2008. *Full Meridian of Glory: Perilous Adventures in the Competition to Measure the Earth*. New York: Springer.

Nagendra, H. 1999. "Mapmakers." *Resonance: Journal of Science Education* 4:8–15.

Needham, J. 1959. *Science and Civilization in China*. Vol. 3, *Mathematics and the Sciences of the Heavens and the Earth*. Cambridge: Cambridge University Press.

Owen, R. 1979. *Great Explorers*. London: Artus Books.

Papenfuse, E. C., and J. M. Coale. 2003. *The Maryland State Archives Atlas of Historical Maps of Maryland, 1608–1908*. Baltimore: Johns Hopkins University Press.

Petrie, G., and T. J. M. Kennie. 1987. "Terrain Modeling in Surveying and Civil Engineering." *Computer Aided Design* 19:171–87.

Price, A. G., ed. 1971. *The Explorations of Captain James Cook in the Pacific*. New York: Dover.
A well-edited version of the great eighteenth-century explorer's three voyages around the world. We expect such explorers to be expert mariners and determined men, but Cook stands out for his surprisingly enlightened views.

Pugh, J. C. 1975. *Surveying for Field Scientists*. London: Methuen.
An excellent overview of pre-GPS techniques.

Quanchi, M., and J. Robson. 2005. *Historical Dictionary of the Discovery and Exploration of the Pacific Islands*. Lanham, MD: Scarecrow Press.
Both a reference on the European discoveries in the Pacific from 1519 to 1876 and a summary of the earlier expansion of South Sea Islanders into these regions.

Raju, P. L. N. 2004. "Fundamentals of Geographical Information System." In *Satellite Remote Sensing and GIS Applications in Agricultural Meteorology*. Geneva, Switzerland: World Meteorological Organization.

Ravenstein, E. G. 1998. *A Journal of the First Voyage of Vasco Da Gama (1497–1499)*. New Delhi: Asian Educational Services.

Rice, E. 1990. *Captain Sir Richard Francis Burton*. New York: Charles Scribner & Sons.
A detailed biography of this odd Victorian traveler: religious mystic, lecher, feminist, rugged explorer of wild regions, writer, linguist, and spy.

Rikitake, T. 1958. "Oscillations of a System of Disk Dynamos." *Proceedings of the Cambridge Philosophical Society* 54:89–105.

Selin, H., ed. 1997. "Physics." *Encyclopaedia of the History of Science, Technology, and Medicine in Non-Western Cultures*. Dordrecht, The Netherlands: Kluwer Academic Publishers.

de Selincourt, A., trans. 2003. *Herodotus: The Histories*. London: Penguin Classics.
The works of the classical historian include accounts of early explorations. This translation includes insightful annotations.

Sheynin, O. 2004. "On the History of the Principle of Least Squares." *Archive for History of Exact Sciences* 46:39–54.

Smith, J. R. 1997. *Introduction to Geodesy*. New York: John Wiley & Sons.

Snyder, J. P. 1993. *Flattening the Earth: Two Thousand Years of Map Projections*. Chicago: University of Chicago Press.

Sobel, D. 1996. *Longitude*. London: Fourth Estate.
Describes John Harrison's lengthy struggles to construct an accurate marine chronometer. Understates the contribution made by French horologists, but tells the story well.

Stearns, B. 1980. "LORAN-C: Low-Cost Electronic Navigation System." *Popular Science*, Aug., 76–78.
A very readable account of LORAN-C use.

Taylor, A. 2004. *The World of Gerard Mercator: The Mapmaker Who Revolutionized Geography*. New York: Walker Publishing.
This biography paints a detailed backdrop, telling us of the times in which the great cartographer lived, and so places the subject in context.

Taylor, E. G. R. 1971. *The Haven-Finding Art*. London: Bodley-Head.

Terrall, M. 2002. *The Man Who Flattened the Earth: Maupertuis and the Sciences in the Enlightenment*. Chicago: University of Chicago Press.
An accessible yet scholarly account of early-eighteenth-century science set in the social context of the times.

Thomas, S. D. 1987. *The Last Navigator*. New York: Henry Holt.

Recounting the accumulated knowledge of a Micronesian navigator from the Caroline Islands.

Thrower, N. J. W., ed. 1984. *Sir Francis Drake and the Famous Voyage, 1577–1580*. Berkeley and Los Angeles: University of California Press.

Toghill, J. 2003. *The Navigator's Handbook*. Guildford, CT: Lyons Press.

Torge, W. 2001. Chapter 1 in *Geodesy*. Berlin: Gruyter.

Turner, A. 1988. *Early Scientific Instruments: Europe, 1400–1800*. London: Philip Wilson Publishers.

Turner, G. L. E. 1998. *Scientific Instruments, 1500–1900: An Introduction*. Berkeley and Los Angeles: University of California Press.

Warkentin, G., ed. 2007. *Canadian Exploration Literature: An Anthology*. Toronto: Dundurn Press.

Weast, R. C., ed. 1973. *CRC Handbook of Physics and Chemistry*. 53rd ed. Cleveland, OH: CRC Press.

Whitaker, R. 2004. *The Mapmaker's Wife*. New York: Random House.

Whitfield, P. 1998. *New Found Lands: Maps in the History of Exploration*. New York: Routledge.

Wilford, J. N. 2000. *The Mapmakers*. Rev. ed. New York: Random House.
An updated and readable history of cartography and geodesy.

Williams, G. E. 1997. "Precambrian Length of Day and the Validity of Tidal Rhythmite Paleotidal Values." *Geophysical Research Letters* 24:421–24.

Wood, F. 2002. Chapter 4 in *The Silk Road*. Berkeley and Los Angeles: University of California Press.

Yang, Q., J. P. Snyder, and W. R. Tobler. 2000. *Map Projection Transformation: Principles and Applications*. London: Taylor and Francis.

Index